MARKOV POINT PROCESSES
—————— AND ——————
THEIR APPLICATIONS

MARKOV POINT PROCESSES
—————— AND ——————
THEIR APPLICATIONS

M. N. M. van Lieshout
Centrum voor Wiskunde en Informatica
The Netherlands

ICP

Imperial College Press

Published by

Imperial College Press
57 Shelton Street
Covent Garden
London WC2H 9HE

Distributed by

World Scientific Publishing Co. Pte. Ltd.
P O Box 128, Farrer Road, Singapore 912805
USA office: Suite 1B, 1060 Main Street, River Edge, NJ 07661
UK office: 57 Shelton Street, Covent Garden, London WC2H 9HE

British Library Cataloguing-in-Publication Data
A catalogue record for this book is available from the British Library.

MARKOV POINT PROCESSES AND THEIR APPLICATIONS

ISBN 1-86094-071-4

This book is printed on acid-free paper.

Printed in Singapore by Uto-Print

In memory of

Nic. van Lieshout

Contents

Chapter 1

Point Processes

1.1 Introduction

Forest data on the location of trees, a sample of cells under a microscope, an object scene to be interpreted by computer vision experts, interacting particles of interest to physicists and a sketch of line segments representing geological faults all share a particular trait: they are presented in the form of a map of geometric objects scattered over some region. Such data by their very nature are spatial, and any analysis has to take this into account. For instance, when the region mapped is large, lack of spatial homogeneity may well result in data that are dense in some regions, while sparse in others. Moreover, interaction between the objects (generically called points in the sequel) influences the appearance of the mapped pattern. In biological applications for example, competition for food or space may cause repulsion between the points; on the other hand, if the observed pattern can be seen as the descendants of some unobserved predecessors, the map may look aggregated due to a clustering of the offspring around their ancestor. However, it is important to realise that apparently clustered patterns may also arise from spatial inhomogeneity rather than interpoint interactions, especially at larger scales.

As a motivating example, suppose we want to implement sustainable farming techniques that apply fertilisers and pesticides only there where it is needed, thus causing minimal damage to the environment [15; 215]. Since pesticides seep to the ground water through pores and cracks in the soil, studying the stain patterns in horizontal soil slices treated with a dye tracer provides useful environmental information [26; 51; 87; 206]. One

1

such pattern, collected by Hatano and Booltink [87] in reclaimed land in the Netherlands and further studied in [87; 206; 207], is given in Figure 1.1. Here, the soil stains are represented by the coordinates of their center of gravity. The stain pattern has a clustered appearance: most points are close to other points, and groups of stains are separated by open spaces that are large in comparison to the typical distance between a point and its nearest neighbour. A closer investigation also reveals repulsion at short range as a result of the non-negligible size of the stains.

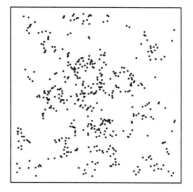

Fig. 1.1   Stain centres in methylene-blue coloured clay observed in a square horizontal slice of side 20 cm sampled at 4 cm depth (Hatano and Booltink, 1992).

A more detailed analysis divides the stains in different types. Thus, elongated stains may be labelled as 'cracks', and roughly circular ones as 'pores'. Those stains that are neither cracks nor pores are called 'vughs'. Distinguishing between these types naturally leads to a trivariate point pattern, the three components representing the cracks, vughs and pores respectively (see Figure 1.2). This approach allows one to investigate each stain type separately as well as to study the correlations between the component patterns. For the data in Figure 1.2, except at short range, both the vugh

and pore components seem aggregated. Moreover vughs and pores tend
to come together, indicating a positive correlation between them. There
are only five cracks, making it hard to say anything meaningful about this
component.

Fig. 1.2   Trivariate point pattern of crack (left), vugh (middle) and pore centres (right)
in methylene-blue coloured clay observed in a square horizontal slice of side 20 cm sampled at 4 cm depth (Hatano and Booltink, 1992).

Finally, the shape of soil stains may be taken into account more explicitly by marking each center of gravity by a shape descriptor. In our
context a useful descriptor is the iso-perimetric ratio defined as the area
of a stain divided by its squared perimeter. In this way, a marked point
pattern is obtained as depicted in Figure 1.3. The stains are represented by
circles around the center with the iso-perimetric ratio as radius. The more
detailed information can be exploited by studying the correlation between
the marks [209].

In this book we will study a flexible class of models for spatial data such
as those discussed above, namely Markov point processes. These models
are defined in terms of local interactions between the points, making them
easy to interpret as well as convenient to work with. This chapter reviews
the fundamentals of general point process theory that are needed in the
remainder of the book. Proofs are included for completeness' sake, but the
less mathematically inclined reader may skip them. Chapter 2 introduces
Markov point processes and investigates their basic properties. The third
chapter is devoted to statistical inference for Markov models, and Chapter 4

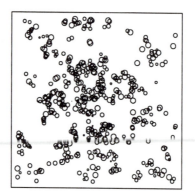

Fig. 1.3   Pattern of soil stains marked by the iso-perimetric ratio in a square horizontal slice of side 20 cm sampled at 4 cm depth (Hatano and Booltink, 1992).

discusses in detail some important examples and fields of application.

1.2   Definitions and notation

Mapped data such as that depicted in Figure 1.1 can be described mathematically as a finite, unordered set of points

$$\mathbf{x} = \{x_1, \ldots, x_n\}, \qquad n = 0, 1, \ldots$$

in some space $\mathcal{X}$ (e.g. a square in $\mathbb{R}^2$). For brevity, such sets will be referred to as *configurations*. Thus, we would like to model the stochastic mechanism underlying the data as a random configuration of objects parameterised by points in $\mathcal{X}$.

In order to give this idea a rigorous measure-theoretical foundation, some structure has to be imposed on the set $\mathcal{X}$. Here, following Daley and Vere–Jones [41], it is assumed that $\mathcal{X}$ is equipped with a metric d such that

$(\mathcal{X}, d)$ is complete and separable*. For example, $\mathcal{X}$ could be a compact subset of $\mathbb{R}^d$ equipped with the Euclidean distance. The metric defines a topology and hence a Borel σ-algebra.

A configuration $\mathbf{x} \subseteq \mathcal{X}$ is said to be *locally finite* if it places at most a finite number of points in any bounded Borel set $A \subseteq \mathcal{X}$. The family of all locally finite configurations will be denoted by $N^{\mathrm{lf}} = N_{\mathcal{X}}^{\mathrm{lf}}$.

Definition 1.1 Let $(\mathcal{X}, d)$ be a complete, separable metric space. A *point process* on $\mathcal{X}$ is a mapping X from a probability space $(\Omega, \mathcal{A}, \mathcal{P})$ into N^{lf} such that for all bounded Borel sets $A \subseteq \mathcal{X}$, the number $N(A) = N_X(A)$ of points falling in A is a (finite) random variable.

From a practical point of view, it is very convenient to work on a metric space. In fact, it may not be feasible to gather data in the form of a mapped pattern, in which case a list of interpoint distances serves as an easier to obtain alternative [40; 46; 219].

Note that the realisations $X(\omega)$, $\omega \in \Omega$, of a point process X are locally finite by definition. Therefore, the cardinality of any $X(\omega)$ is at most countably infinite and there are no accumulation points. If the space $\mathcal{X}$ itself is bounded, or $N_X(\mathcal{X})$ is almost surely finite, the point process X is said to be *finite*.

Example 1.1 Let $\mathcal{X} = [0, a]$ be an interval of length $a > 0$. Define $X = \{U\}$, where the random variable U is uniformly distributed on $[0, a]$. The point process X is well-defined, as $N(A) = \mathbb{I}\{U \in A\}$ is a random variable for any bounded Borel set $A \subseteq \mathcal{X}$. Here we write $\mathbb{I}\{U \in A\}$ for the indicator random variable which takes the value 1 if U falls in A and 0 otherwise. Having only one point, clearly X is finite.

Example 1.2 Let the random variables $X_1, X_2, \ldots$ be independent and exponentially distributed with parameter $\lambda > 0$. Then a *Poisson process* [106] of rate λ on $\mathbb{R}^+$ is defined by

$$X = \{X_1, X_1 + X_2, X_1 + X_2 + X_3, \ldots\}$$

---

*Some authors [40; 98; 100; 130], prefer to work in the more general context of a locally compact, second countable Hausdorff space.

as illustrated in Figure 1.4.

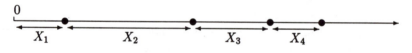

Fig. 1.4   A Poisson process on the positive half line.

Note that for any bounded Borel set A,

$$N(A) = \sum_{k=1}^{\infty} \mathbb{I}\{X_1 + \cdots + X_k \in A\}$$

is a limit of random variables, hence a random variable itself. To show that the number of points in A is almost surely finite, let $(0, t]$ be an interval containing A. Then $N(A) \leq N((0, t])$ almost surely, and it suffices to show that $N((0, t])$ is almost surely finite. Now, since the sum of independent, exponentially distributed random variables has a gamma distribution,

$$\mathbf{P}(N((0, t]) = k) = \mathbf{P}(X_1 + \cdots + X_k \leq t; X_1 + \cdots + X_{k+1} > t)$$

$$= \int_0^t \frac{\lambda^k}{(k-1)!} r^{k-1} e^{-\lambda r} e^{-\lambda(t-r)} \, dr = e^{-\lambda t} \frac{(\lambda t)^k}{k!}$$

for $k \in \mathbb{N}_0$. Hence $N((0, t])$ is Poisson distributed, and in particular almost surely finite.

According to Definition 1.1, a point process X is a random configuration of points such that for each bounded Borel set $A \subseteq \mathcal{X}$, the number of points in A is a random variable. In other words, a point process is a random variable with values in the measurable space $(N^{\mathrm{lf}}, \mathcal{N}^{\mathrm{lf}})$, where $\mathcal{N}^{\mathrm{lf}}$ is the smallest σ-algebra such that for all bounded Borel sets $A \subseteq \mathcal{X}$ the mapping $\mathbf{x} \mapsto N_{\mathbf{x}}(A)$ is measurable [208]. The induced probability measure on $\mathcal{N}^{\mathrm{lf}}$ is called the *distribution* of X.

Other definitions are encountered in the literature. Perhaps the most popular one is in terms of random counting measures putting mass on the points of the process [39; 40; 41; 98; 100; 111; 131; 168]. Alternatively, point patterns can be treated as special cases of random sets [130; 195], or one may work directly with the family $\{N(A)\}_A$ counting the number of points in Borel sets $A \subseteq \mathcal{X}$ [176; 179]. The latter definition requires conditions to ensure for instance that almost surely $N(A \cup B) = N(A) +$

$N(B)$ whenever A and B are disjoint [41]. A measure theoretic approach avoiding the topological structure on $\mathcal{X}$ imposed here can be found in [171; 170; 172].

Since $\mathcal{N}^{lf}$ is defined by the measurability of $\mathbf{x} \mapsto N_{\mathbf{x}}(A)$ for bounded Borel sets $A \subseteq \mathcal{X}$, the analogue of the distribution function for random variables in a point process context is the ensemble of joint distribution functions of vectors $(N(A_1), \ldots, N(A_m))$ where A_i $(i = 1, \ldots, m)$ are bounded Borel sets.

Definition 1.2   The family of *finite-dimensional distributions* (or 'fidis') of a point process X on a complete, separable metric space $(\mathcal{X}, d)$ is the collection of joint distributions of $(N(A_1), \ldots, N(A_m))$ for all finite vectors $(A_1, \ldots, A_m)$ of bounded Borel sets $A_i \subseteq \mathcal{X}$, $i = 1, \ldots, m$, of any length $m \in \mathbb{N}$.

The following characterisation theorem holds [41].

Theorem 1.1   *The distribution of a point process X on a complete, separable metric space $(\mathcal{X}, d)$ is completely specified by its finite-dimensional distributions.*

Thus, if two point processes have the same fidis, they necessarily share the same distribution.

Proof.   The sets

$$\left\{ \omega \in \Omega : N_{X(\omega)}(A_i) \in B_i, i = 1, \ldots, m \right\}$$

for bounded Borel sets $A_i \subseteq \mathcal{X}$ and Borel sets $B_i \subseteq \mathbb{R}$ form a semiring generating $\mathcal{N}^{lf}$. The probabilities of all such sets can be determined from the fidis, hence, using a classical result from real analysis [79], the distribution of X is determined uniquely by its fidis. $\square$

In the agricultural example discussed in section 1.1, for each soil stain its location as well as its iso-perimetric ratio was recorded. Data of this kind are conveniently described by means of *marked point processes* [41]. .

Definition 1.3 Let $(\mathcal{X}, d)$ and $(\mathcal{K}, d')$ be complete, separable metric spaces. A marked point process with positions in $\mathcal{X}$ and marks in $\mathcal{K}$ is a point process on $\mathcal{X} \times \mathcal{K}$ such that the process of unmarked points is a well-defined point process.

The Cartesian product $\mathcal{X} \times \mathcal{K}$ is a complete, separable metric space with respect to the metric

$$\rho((x, k), (y, l)) = \max\{d(x, y), d'(k, l)\}$$

so that definition 1.3 is in keeping with our general set-up.

If the mark space $\mathcal{K} = \{1, \ldots, M\}$ is a finite collection of labels, equipped e.g. with the metric $d'(k, l) = |k - l|$, any point process Y on the product space $\mathcal{X} \times \mathcal{K}$ gives rise to a well-defined process X of unmarked points. To see this, let $A \subseteq \mathcal{X}$ be a bounded Borel set. Then the number of unmarked points in A can be written as

$$N_X(A) = \sum_{i=1}^{M} N_Y(A \times \{i\})$$

where $N_Y(A \times \{i\})$ denotes the number of points in A having label i ($i = 1, \ldots, M$). By definition, each $N_Y(A \times \{i\})$ is a finite random variable, hence $N_X(A)$ is a finite random variable as well. If $\mathcal{K}$ is continuous, more care is needed. For instance a unit rate Poisson process on $\mathbb{R}^3$ (which will be treated in detail in section 1.5) is not a marked point process on $\mathbb{R}^2$ with marks in $\mathbb{R}$. Indeed,

$$N_X(A) = N_Y(A \times \mathbb{R})$$

is not necessarily finite for all bounded Borel sets $A \subseteq \mathbb{R}^2$.

Example 1.3 A multivariate point process

$$Y = (X_1, \ldots, X_M)$$

can be seen as a marked point process with marks denoting the component labels. Since the mark space $\{1, \ldots, M\}$ is finite, the process $X = \cup_{i=1}^{M} X_i$ of unmarked points is well-defined.

1.3 Simple point processes

In most practical examples, a mapped point pattern will not contain multiple points at exactly the same place, either because it is physically impossible as in the soil example of section 1.1, or since the multiplicity is captured in a label attached to each observed location. An example of the former case is when points represent the center of a rigid body such as a tree or a non-penetrable cell; the latter situation arises for instance in rare animal sightings, where the spot of a sighting is recorded and any returns of the animal are either deemed to belong to the same event or registered as a mark.

More formally, let N_s^{lf} denote the set of all locally finite point configurations $\mathbf{x}$ consisting of distinct points, i.e. $N_{\mathbf{x}}(\{x\}) \in \{0,1\}$ for all $x \in \mathcal{X}$. To verify that N_s^{lf} is $\mathcal{N}^{\text{lf}}$–measurable, note that as $\mathcal{X}$ is separable it can be covered by a countable number of open balls $S(x_i, \left(\frac{1}{2}\right)^j)$ of arbitrary small radius. Consequently $N_s^{\text{lf}} = \bigcap_{i=1}^{\infty} \left\{ \omega \in \Omega : N\left(S\left(x_i, \left(\frac{1}{2}\right)^j\right)\right) \in \{0,1\} \right\} \in \mathcal{N}^{\text{lf}}$.

Definition 1.4 A point process X is simple if it takes its values in N_s^{lf} almost surely.

Simple point processes are particularly appealing to work with, since we do not need the full family of fidis to specify their distribution. Indeed, it is sufficient to know the *void probabilities*

$$v(A) = \mathbf{P}(N(A) = 0) \tag{1.1}$$

for a sufficiently large class of sets $A \subseteq \mathcal{X}$ [41; 131; 133; 139; 169].

Theorem 1.2 *The distribution of a simple point process X on a complete, separable metric space $(\mathcal{X}, d)$ is uniquely determined by the void probabilities of bounded Borel sets $A \subseteq \mathcal{X}$.*

In the above theorem, the collection of bounded Borel sets may be replaced by a smaller class of test sets such as the compact sets. For details see [36; 41; 131; 133]. The proof [41] of theorem 1.2 is rather technical and may be omitted on first reading.

Proof. By Theorem 1.1 it suffices to show that the fidis

$$\mathbf{P}(N(A_1) \le n_1; \ldots; N(A_k) \le n_k)$$

for $k \in \mathbb{N}$, bounded Borel sets $A_1, \ldots A_k$, and non-negative integers $n_1, \ldots, n_k$ are completely determined by the void probabilities (1.1).

To do so, define a family of difference operators [36] $S_k(\cdot\ ; A_1, \ldots, A_k)$ indexed by bounded Borel sets inductively as follows:

$$S_1(B; A_1) = v(B) - v(A_1 \cup B)$$
$$S_k(B; A_1, \ldots, A_k) = S_{k-1}(B; A_1, \ldots, A_{k-1}) - S_{k-1}(B \cup A_k; A_1, \ldots A_{k-1})$$

where $B \subseteq \mathcal{X}$ is a bounded Borel set. The operators may be given a probabilistic interpretation by noting that $S_1(B; A_1) = \mathbf{P}(N(B) = 0) - \mathbf{P}(N(A_1 \cup B) = 0) = \mathbf{P}(N(A_1) > 0; N(B) = 0)$. Similarly

$$S_k(B; A_1, \ldots, A_k) = \mathbf{P}(N(A_i) > 0, i = 1, \ldots, k; N(B) = 0). \qquad (1.2)$$

The family of difference operators $(S_k)_k$ and therefore all probabilities of the form (1.2) are completely determined by the void probabilities.

The remainder of the proof is devoted to approximating the fidis using a collection of random variables $N(A_{ni})$ with values in $\{0, 1\}$ only. The idea is that if the test sets A_{ni} are 'small' enough, they contain at most a single point of the process.

Let $\{T_{ni} : i = 0, \ldots k_n\}$ $(n \in \mathbb{N})$ be a nested sequence of partitions of $\mathcal{X}$ defined as follows. Since $\mathcal{X}$ is separable, there exists a countable dense set $\mathcal{D} = \{d_1, d_2, \ldots\}$. The first partition is simply

$$T_{11} = S(d_1, 1), \quad T_{10} = \mathcal{X} \setminus T_{11},$$

where $S(d_1, 1)$ denotes the open ball of radius 1 centered at d_1. Inductively, having defined nested partitions up to level $n - 1$, define for $i = 1, \ldots, n$ the sets $B_{ni} = S(d_i, \frac{1}{2}^{n-i})$ and $B_{n0} = \mathcal{X} \setminus \cup_{i=1}^n B_{ni}$; to obtain a partition, set $C_{n0} = B_{n0}$, $C_{n1} = B_{n1}$ and $C_{nj} = B_{nj} \setminus \cup_{i=1}^{j-1} B_{nj}$. The intersections $T_{n-1,j} \cap C_{nk}$ with the partition at level $n-1$ then form a partition $(T_{ni})_i$ that is nested in $(T_{n-1,i})_i$. Note that, eventually, distinct points are separated, that is, belong to different members of the partition. The procedure is illustrated in Figure 1.5 for $n = 2$.

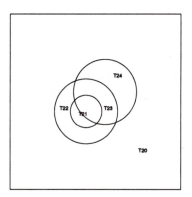

Fig. 1.5   The nested partition $(T_{2i})_i$ at level 2.

Returning to the fidis, consider the limit

$$\lim_{n\to\infty} H_n(A_1) = \lim_{n\to\infty} \sum_{i=0}^{k_n} H(A_{ni}),$$

where $A_{ni} = T_{ni} \cap A_1$ and $H(A_{ni}) = \mathbb{I}\{N(A_{ni}) \geq 1\}$. Note that the family $\{A_{ni} = T_{ni} \cap A_1\}_{n,i}$ is a nested partition of A_1 that separates distinct points; the random variable $H_n(A_1)$ counts the number of sets in the nth partition restricted to A_1 containing points of X. Since the point process X is simple, and $H_n(A_1)$ is monotonically increasing,

$$\lim_{n\to\infty} H_n(A_1) = N(A_1)$$

exists almost surely. To show that the joint distribution of $H(A_{ni})$ can be expressed in terms of the difference operator (1.2) and hence is completely specified by the void probabilities, observe that since each $H(A_{ni})$ is an indicator random variable taking binary values,

$$\mathbf{P}(H(A_{n0}) = i_0; \cdots; H(A_{nk_n}) = i_{k_n}) = S_l(\cup_{j:i_j=0} A_{nj}; A_{nj}, j : i_j = 1)$$

(writing l for the number of indices j for which $i_j = 1$). Hence

$$\mathbf{P}(H_n(A_1) = l) = \sum \mathbf{P}(H(A_{n0}) = i_0; \cdots ; H(A_{nk_n}) = i_{k_n})$$

with the sum taken over all combinations of i_js $\in \{0, 1\}$ summing to $l \in \mathbb{N}_0$. Therefore, $\mathbf{P}(H_n(A_1) = l)$ is expressed solely in terms of $v(\cdot)$. More generally, the joint distribution of $H_n(A_1), \ldots, H_n(A_k)$ for $k \in \mathbb{N}$ can be written in terms of the void probabilities. Finally, the event $\{H_n(A_1) \leq n_1; \cdots ; H_n(A_k) \leq n_k\}$ is monotonically decreasing in n, hence

$$\mathbf{P}(H_n(A_1) \leq n_1; \cdots ; H_n(A_k) \leq n_k) \to \mathbf{P}(N(A_1) \leq n_1; \cdots ; N(A_k) \leq n_k)$$

as $n \to \infty$. $\qquad\qquad\qquad\qquad\qquad\qquad\qquad\qquad\qquad\qquad\qquad\square$

Example 1.4   The uniform random point considered in example 1.1 is simple. In order to find its void probabilities, note that $\mathbf{P}(u \in A) = \frac{l(A)}{a}$ is proportional to the 'length' (Lebesgue measure) $l(A)$ of A, and hence

$$v(A) = \mathbf{P}(U \in \mathcal{X} \setminus A) = \frac{l(\mathcal{X} \setminus A)}{a}.$$

Example 1.5   The 1–dimensional Poisson process of example 1.2 is simple, as the inter event distances $X_1, X_2, \ldots$ (cf. Figure 1.4) are almost surely positive. Furthermore, for any $0 < s < t$,

$$
\begin{aligned}
v((s, t]) &= \mathbf{P}(X_1 > t) + \sum_{k=1}^{\infty} \mathbf{P}(X_1 + \cdots + X_k \leq s; X_1 + \cdots + X_{k+1} > t) \\
&= e^{-\lambda t} + \sum_{k=1}^{\infty} \int_0^s \frac{\lambda^k}{(k-1)!} r^{k-1} e^{-\lambda r} e^{-\lambda(t-r)} \, dr \\
&= \sum_{k=0}^{\infty} \frac{(\lambda s)^k}{k!} e^{-\lambda t} = \exp[-\lambda(t - s)].
\end{aligned}
$$

Let us briefly consider two point processes X and Y that are not simple but do have the same void probabilities. Then the processes X^s and Y^s obtained from X and Y respectively by 'ignoring the multiplicities' are simple and have identical void probabilities. Therefore, by Theorem 1.2 the distributions of X^s and Y^s are the same, hence the distributions of X and Y differ only with respect to the multiplicity of the points.

1.4 Finite point processes

Apart from being simple (cf. section 1.3), most point patterns encountered in practice are observed in a bounded region. Sometimes this region is dictated by the application; more often the spatial process of interest extends over a space that is too large to be mapped exhaustively and data are recorded in a smaller 'window' chosen for convenience. In any case, the resulting map contains a *finite* number of points.

A convenient and constructive way [41] to model finite patterns is by means of

- a discrete probability distribution $(p_n)_{n \in \mathbb{N}_0}$ for the number of points;
- a family of symmetric probability densities $j_n(x_1, \ldots, x_n)$, $n \in \mathbb{N}$, on $\mathcal{X}^n$ for the locations.

Here we assume that $\mathcal{X}$ is equipped with a Borel measure $\nu(\cdot)$, so that densities $j_n(\cdot, \ldots, \cdot)$ can be defined with respect to the product measure $\nu(\cdot)^n$. Thus, a point process X may be constructed as follows. Let $N(\mathcal{X})$ be a random variable with distribution $(p_n)_n$, and, conditionally on $\{N(\mathcal{X}) = n\}$, let $(X_1, \ldots, X_n) \in \mathcal{X}^n$ be a random vector distributed according to $j_n(\cdot, \ldots, \cdot)$ independently of $N(\mathcal{X})$. The symmetry requirement for the location densities is needed, since a point process is indifferent with respect to the order in which its points are listed.

A conscientious reader may wonder whether the implicit transition from ordered vectors to unordered configurations is allowed. To see that this is indeed the case [41; 168], first note that $N_n^f = \{\mathbf{x} \in N^f : N_{\mathbf{x}}(\mathcal{X}) = n\}$ is a measurable subset of N^f, the collection of finite point configurations, with respect to the smallest σ-algebra $\mathcal{N}^f$ for which the mappings $\mathbf{x} \mapsto N_{\mathbf{x}}(A)$ (A bounded Borel) are measurable. Thus, we may define a σ-algebra on N_n^f by the trace $\mathcal{N}_n^f$ of $\mathcal{N}^f$. Write $f_n : \mathcal{X}^n \to N_n^f$ for the function that maps n-vectors to configurations of n points. Then f_n is Borel-measurable, and, because of its permutation invariance, measurable with respect to the σ-algebra $\mathcal{B}_s(\mathcal{X}^n)$ of symmetric Borel sets in $\mathcal{X}^n$ as well.

By assumption $j_n(\cdot, \ldots, \cdot)$ is a permutation invariant density for each $n \in \mathbb{N}$, and therefore $\mathcal{B}_s(\mathcal{X}^n)$-measurable. Composition with f_n yields a function $i_n : N_n^f \to \mathbb{R}$ on configurations of n points:

$$i_n(\mathbf{x}) = \frac{1}{n!} \sum_\varphi j_n(\varphi(\mathbf{x}))$$

with the sum ranging over all permutations of $\mathbf{x}$. Clearly, $i_n \circ f_n = j_n$. To show that i_n is measurable, let $A \subseteq \mathcal{X}^n$ be a bouded Borel set. We have to verify that $i_n^{-1}(A) \in N_n^f$. Now, as f_n is surjective, $i_n^{-1}(A) = f_n\left(f_n^{-1}(i_n^{-1}(A))\right) = f_n\left(j_n^{-1}(A)\right)$. Since j_n is measurable with respect to the σ-algebra of symmetric Borel sets, it suffices to show that $f_n(B) \in N_n^f$ for all $B \in \mathcal{B}_s(\mathcal{X}^n)$. Define the σ-algebra $\mathcal{A}_n = \left\{ B \in \mathcal{B}_s(\mathcal{X}^n) : f_n(B) \in N_n^f \right\}$. Then $\mathcal{A}_n$ includes all rectangles A^n with A bounded Borel, and, since such rectangles generate $\mathcal{B}_s(\mathcal{X}^n)$, it follows that $\mathcal{A}_n = \mathcal{B}_s(\mathcal{X}^n)$. Consequently, $f_n(B) \in N_n^f$ for all $B \in \mathcal{B}_s(\mathcal{X}^n)$, and there is a 1-1 correspondence between functions on the configuration space N_n^f and symmetric functions on $\mathcal{X}^n$.

Fig. 1.6   Realisation of a binomial process of 100 points in the unit square.

Example 1.6   Let $\mathcal{X}$ be a compact subset of $\mathbb{R}^d$ of strictly positive volume

$\mu(\mathcal{X})$. A *binomial point process* [100; 168; 208] is defined as the union

$$X = \{X_1, \ldots, X_n\}$$

of a fixed number $n \in \mathbb{N}$ of independent, uniformly distributed points $X_1, \cdots, X_n$.

Since $\mathbf{P}(X_i = X_j) = 0$ for all $i \neq j$, X is simple. Furthermore, as $\mathbf{P}(N(\mathcal{X}) = n) = 1$, the binomial process is finite with

$$p_m = \begin{cases} 0 & \text{if } m \neq n \\ 1 & \text{if } m = n \end{cases}.$$

The points X_i are distributed uniformly, hence

$$j_n(x_1, \ldots, x_n) \equiv \left(\frac{1}{\mu(\mathcal{X})} \right)^n$$

$(x_1, \ldots, x_m \in \mathcal{X})$. Clearly, $j_n(\cdot, \ldots, \cdot)$ is permutation invariant.

The binomial process derives its name from the fact that for any Borel set $A \subseteq \mathcal{X}$,

$$N(A) = \sum_{i=1}^{n} \mathbb{I}\{X_i \in A\}$$

follows a binomial distribution with parameters n and $\mu(A)/\mu(\mathcal{X})$. A simulation for $n = 100$ and $\mathcal{X} = [0,1]^2$ is given in Figure 1.6.

Example 1.7   Further to example 1.6, if the d-volume of $\mathcal{X}$ is zero it may be possible to identify $\mathcal{X}$ with a lower dimensional set $\mathcal{Y} \subseteq \mathbb{R}^c$ $(c < d)$ of positive c-volume. For example, points on the boundary of the unit sphere in $\mathbb{R}^2$ can be identified by a single orientation parameter. Thus, if $X_1, \ldots, X_n$ are independent, uniformly distributed 'angles' in $[0, 2\pi)$,

$$X = \{(\cos X_1, \sin X_1), \cdots, (\cos X_n, \sin X_n)\}$$

is a binomial process on the unit sphere.

1.5   Poisson point processes

Perhaps the best-known and most tractable of point processes is the Poisson model [39; 40; 41; 46; 100; 106; 131; 130; 168; 176; 179; 195; 208; 209]. Its interest lies predominantly in the fact that it represents spatial randomness, a notion made more precise in theorem 1.3 below. Indeed, the first step in analysing a point pattern often is to test for spatial randomness. Rejection of the Poisson model may yield valuable insight in the pattern at hand and suggest alternative classes of models. For more details, see sections 1.8, 4.8 or [46].

Below, we shall derive a Poisson process on $\mathbb{R}^d$ in an intuitive way [208] from the binomial process discussed in example 1.6. The results will serve as the basis for a formal definition.

Let then, for $n \in \mathbb{N}$, $P^{(n)}(\cdot)$ be the distribution of a binomial process of n points in a ball $B_n \subseteq \mathbb{R}^d$ centred at the origin, with radius chosen in such a way that the volume $\mu(B_n) = \frac{n}{\lambda}$ for some $0 < \lambda < \infty$. Furthermore, let A be a bounded Borel set, and k a non-negative integer. Since the sequence of balls is increasing towards $\mathbb{R}^d$, an integer number $n_0 \geq k$ can be found such that for $n \geq n_0$, $A \subseteq B_n$. Hence, for $n \geq n_0$,

$$\begin{aligned}
\mathbf{P}^{(n)}(N(A) = k) &= \mathbf{P}^{(n)}(N(A) = k; N(B_n \setminus A) = n - k) \\
&= \binom{n}{k} \left(\frac{\mu(A)}{\mu(B_n)} \right)^k \left(\frac{\mu(B_n \setminus A)}{\mu(B_n)} \right)^{n-k}.
\end{aligned}$$

Hence the number of points in A is binomially distributed with parameters n and $\mu(A)/\mu(B_n)$. Hence

$$\lim_{n \to \infty} \mathbf{P}^{(n)}(N(A) = k) = e^{-\lambda \mu(A)} \frac{(\lambda \mu(A))^k}{k!}.$$

Further, if A and B are disjoint bounded Borel sets, there exists an $n_0 \in \mathbb{N}$ such that $A \cup B \subseteq B_n$ for $n \geq n_0$. Then, for all fixed $k, l \in \mathbb{N}_0$ and all $n \geq \max\{n_0, k + l\}$, the joint probability $\mathbf{P}^{(n)}(N(A) = k; N(B) = l) = \mathbf{P}^{(n)}(N(A) = k; N(B) = l; N(B_n \setminus (A \cup B)) = n - k - l$ equals

$$\binom{n}{k} \left(\frac{\mu(A)}{\mu(B_n)} \right)^k \binom{n-k}{l} \left(\frac{\mu(B)}{\mu(B_n)} \right)^l \left(\frac{\mu(B_n \setminus (A \cup B))}{\mu(B_n)} \right)^{n-k-l} =$$

$$\binom{n}{k} \left(\frac{\mu(A)}{\mu(B_n)} \right)^k \left(\frac{\mu(B_n \setminus A)}{\mu(B_n)} \right)^{n-k} \binom{n-k}{l} \left(\frac{\mu(B)}{\mu(B_n)} \right)^l \left(\frac{\mu(B_n \setminus B)}{\mu(B_n)} \right)^{n-k-l}$$

$$\left(\frac{\mu(B_n \setminus (A \cup B))}{\mu(B_n)} \right)^{n-k-l} \left(\frac{\mu(B_n)}{\mu(B_n \setminus A)} \right)^{n-k} \left(\frac{\mu(B_n)}{\mu(B_n \setminus B)} \right)^{n-k-l}.$$

Since the sets A and B as well as the integer numbers k and l are fixed, and $n/\mu(B_n) \to \lambda$ by assumption,

$$\left(\frac{\mu(B_n \setminus (A \cup B))}{\mu(B_n)}\right)^{n-k-l} \to e^{-\lambda\mu(A \cup B)}$$

as $n \to \infty$. Similarly, $(\frac{\mu(B_n)}{\mu(B_n \setminus A)})^{n-k} \to e^{\lambda\mu(A)}$ and $(\frac{\mu(B_n)}{\mu(B_n \setminus B)})^{n-k-l} \to e^{\lambda\mu(B)}$. We conclude that in the limit $N(A)$ and $N(B)$ are independent and Poisson distributed, with parameters $\lambda\mu(A)$ and $\lambda\mu(B)$ respectively.

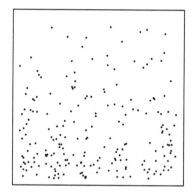

Fig. 1.7 Realisation of an inhomogeneous Poisson process on the unit square with intensity function $\lambda(x, y) = 600e^{-3y}$.

The properties derived above suggest the following definition. A point process X on $\mathcal{X} = \mathbb{R}^d$ is a *homogeneous Poisson process* with *intensity* (or *rate*) $\lambda > 0$ if

- $N(A)$ is Poisson distributed with mean $\lambda\mu(A)$ for every bounded Borel set $A \subseteq \mathcal{X}$, writing $\mu(A)$ for the volume (Lebesgue measure) of A;
- for any k disjoint bounded Borel sets $A_1, \ldots, A_k$, the random variables $N(A_1), \ldots, N(A_k)$ are independent.

Various generalisations spring to mind. For instance, the space $\mathbb{R}^d$ can be replaced by any compact set $\mathcal{X}$ of positive volume $\mu(\mathcal{X}) > 0$. Furthermore, an *inhomogeneous Poisson process* may be defined by the requirement that the number of points in a bounded Borel set A has a Poisson distribution with parameter

$$\int_A \lambda(x)\, dx < \infty \tag{1.3}$$

where $\lambda(\cdot)$ is any non-negative, Borel measurable intensity function on $\mathcal{X}$. As an illustration consider Figure 1.7, showing a realisation of a Poisson process on $[0,1]^2$ with $\lambda(x,y) = \mu\lambda e^{-\lambda y}$ ($\mu = 200$, $\lambda = 3$). Thus, the intensity is constant along the x-axis and decays exponentially in y. The relative abundance of points in the lower part of the square is clearly visible in the map (cf. Figure 1.7).

We are now ready to state the definition of a Poisson process on an arbitrary complete separable metric space $(\mathcal{X}, d)$.

Definition 1.5   Let $\nu(\cdot)$ be a Borel measure on a complete separable metric space $(\mathcal{X}, d)$ such that $\nu(\mathcal{X}) > 0$ and $\nu(A) < \infty$ for all bounded Borel sets A. Measures satisfying the latter property are called *locally finite*. A point process X on $\mathcal{X}$ is a *Poisson process* with intensity measure $\nu(\cdot)$ if

(P1) $N(A)$ is Poisson distributed with mean $\nu(A)$ for every bounded Borel set $A \subseteq \mathcal{X}$;

(P2) for any k disjoint bounded Borel sets $A_1, \dots, A_k$, the random variables $N(A_1), \dots, N(A_k)$ are independent.

If $\nu(\cdot)$ is atomless, the Poisson process is simple (cf. definition 1.4); it is finite if $\nu(\mathcal{X}) < \infty$.

Property (P2) can be interpreted as *complete spatial randomness* in the sense that point configurations in disjoint regions behave independently of one another. From a practical point of view, (P2) implies that observing a point process in an arbitrarily chosen window W only does not introduce biases or edge effects since the pattern outside of the window does not affect the behaviour seen in W. In other words, (P2) formalises a lack of interdependence between disjoint regions.

The theorem below states that the points of a Poisson process also behave randomly and do not interact with each other [41; 46; 100; 168; 208].

Theorem 1.3 *Let X be a Poisson process on a complete separable metric space $(\mathcal{X}, d)$ with intensity measure $\nu(\cdot)$, and $A \subseteq \mathcal{X}$ a bounded Borel set. Then, conditional on $\{N(A) = n\}$, X restricted to A is distributed as a binomial process of independent, ν-uniformly distributed points.*

Proof. Pick any Borel set $B \subseteq A$. The conditional void probability of B given n points in A is

$$v_A(B) = \mathbf{P}(N(B) = 0 \mid N(A) = n) = \frac{\mathbf{P}(N(B) = 0; N(A \setminus B) = n)}{\mathbf{P}(N(A) = n)}.$$

By (P2), $N(B)$ and $N(A \setminus B)$ are independent. Furthermore $N(A), N(B)$ and $N(A \setminus B)$ are Poisson distributed by (P1). Hence

$$v_A(B) = \left(\frac{\nu(A \setminus B)}{\nu(A)} \right)^n,$$

coinciding with the void probability of B for n independent ν-uniform points (see example 1.1). Since B is arbitrary, an application of theorem 1.2 completes the proof. □

Theorem 1.3 above was used to obtain the sample pictured in Figure 1.7. First, a realisation of a Poisson random variable N with mean $\mu(1 - e^{-\lambda})$ was generated to determine the total number of points. Conditionally on $N = n$, the n points were simulated independently. In our case, since the intensity function is constant along the x-axis, the first coordinate is uniformly distributed; the second coordinate has density function $\frac{\lambda e^{-\lambda y}}{1 - e^{-\lambda}}$ on $[0, 1]$ independently of the x-coordinate.

The following examples deal with the simulation of Poisson processes in cases where the intensity function or the space is intractable. Both examples are special cases of *rejection sampling* [178].

Example 1.8 Let X be an inhomogeneous Poisson process on $\mathbb{R}^d$ with bounded intensity function $\lambda(\cdot) \leq \Lambda$. When $\lambda(\cdot)$ is too complicated to sample from directly, the following observation often helps [115]. Suppose that X_h is a homogeneous Poisson process of intensity Λ and that each of its points is deleted independently of other points with a location dependent probability. More precisely, a point at $x \in \mathcal{X}$ is retained with probability $\frac{\lambda(x)}{\Lambda}$. Then the void probabilities $v_t(A)$, A bounded Borel, of the thinned process are

$$
\begin{aligned}
v_t(A) &= \sum_{k=0}^{\infty} \mathbf{P}(N(X_h \cap A) = k; \text{ all points of } X_h \cap A \text{ are deleted }) \\
&= \sum_{k=0}^{\infty} \left[e^{-\Lambda\mu(A)} \frac{\Lambda^k \mu(A)^k}{k!} \left(\int_A \frac{1}{\mu(A)} \left(1 - \frac{\lambda(x)}{\Lambda} \right) dx \right)^k \right] \\
&= \sum_{k=0}^{\infty} e^{-\Lambda\mu(A)} \frac{\Lambda^k \mu(A)^k}{k!} \left(1 - \frac{\nu(A)}{\Lambda\mu(A)} \right)^k = e^{-\nu(A)} \qquad (1.4)
\end{aligned}
$$

where $\nu(A) = \int_A \lambda(x)\, dx$. Since the process X of interest places a Poisson number of points in A with mean $\nu(A)$, its void probabilities are identical to (1.4). Hence by theorem 1.2, a sample from X can be obtained simply by thinning a realisation of a homogeneous Poisson process!

Example 1.9   A Poisson process in an irregularly shaped set A (see Figure 1.8) can be obtained by simply choosing a more tractable set $\mathcal{X} \supseteq A$, generating a realisation of the process in $\mathcal{X}$, and deleting all points that do not fall in A.

From the preceding examples it is clear that the conditional approach suggested by theorem 1.3 is a useful simulation technique. However, it is not the only one. For instance [115], realisations of a homogeneous Poisson process in a square $[0, a] \times [0, b]$ can also be obtained by first generating the x-coordinates through a series of exponential increments, followed by sampling the y-coordinates uniformly on $[0, b]$.

As well as suggesting simulation algorithms, theorem 1.3 also provides a way to *construct* the actual Poisson process itself. For instance, if $\mathcal{X} \subset \mathbb{R}^d$ is a compact Borel set, let N be a Poisson distributed random variable with mean $\lambda\mu(\mathcal{X})$ and suppose that X_i ($i \in \mathbb{N}$) is a sequence of independent,

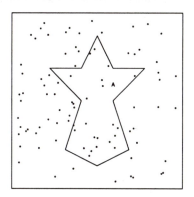

Fig. 1.8 Realisation of a Poisson process of intensity 100 in $A \subseteq [0,1]^2$.

uniformly distributed points in $\mathcal{X}$. Set

$$X = \bigcup_{i=1}^{N} X_i.$$

Conditioning over the outcomes of N, by arguments similar to those used in the heuristic derivation preceding definition 1.5, it can be shown that X is a well-defined point process satisfying properties (P1)–(P2). It follows that definition 1.5 is not vacuous, i.e., that homogeneous Poisson processes exist.

More generally, a Poisson process with intensity measure $\nu(\cdot)$ on a complete, separable metric space $\mathcal{X}$ can be constructed by partitioning $\mathcal{X}$ into bounded subsets C_i of positive measure $\nu(C_i)$. For each i independently, one defines a process $X^{(i)}$ as above, and takes their union to form the desired Poisson process. For details see [168].

The next example shows that theorem 1.3 is useful for computations.

Example 1.10 Let us compute the average number of pairs of points in a Poisson process of intensity λ on the planar unit square separated by a distance that does not exceed some fixed $r < \sqrt{2}$. Note that this pair count is permutation invariant, hence measurable as a function of configurations (cf. the discussion in section 1.4).

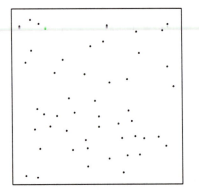

Fig. 1.9 Realisation of a dependent thinning of a Poisson process of rate 100 in the unit square. The hard core radius is 0.05.

Conditionally on $N([0,1]^2) = n$, the points are uniformly distributed. Therefore

$$\mathbf{E}\left[\sum_{i=1}^{n}\sum_{j=i+1}^{n}\mathbb{I}\{\|X_i - X_j\| \leq r\} \,\big|\, N([0,1]^2) = n\right]$$

equals

$$\frac{1}{2}\,n(n-1)\int_{[0,1]^2}\int_{[0,1]^2}\mathbb{I}\{\|x_1 - x_2\| \leq r\}\,dx_1\,dx_2.$$

Since $N([0,1]^2)$ is Poisson distributed with parameter λ, the unconditional expectation of the number of r-close pairs equals

$$\frac{1}{2}\lambda^2 \int \int_{[0,1]^2} \mathbb{I}\{\|x_1 - x_2\| \leq r\} \, dx_1 \, dx_2.$$

A point process in which pairs of r-close points are prohibited is called a *hard core process*. An example, obtained by deleting all r-close pairs from a realisation of a Poisson process, is shown in Figure 1.9. For more examples of marked and thinned Poisson processes, see [220].

To conclude this section, we present the point process analogue of the well-known fact that the sum of two independent Poisson random variables is again Poisson distributed [131].

Theorem 1.4 *Let X and Y be independent Poisson processes on the complete separable metric space $(\mathcal{X}, d)$ with non-atomic intensity measures $\lambda(\cdot)$ and $\nu(\cdot)$ respectively. Then the superposition $Z = X \cup Y$ is a Poisson process with intensity measure $\lambda(\cdot) + \nu(\cdot)$.*

Proof. The probability that a bounded Borel set $A \subseteq \mathcal{X}$ contains no points of Z is

$$
\begin{aligned}
\mathbf{P}(X \cap A = \emptyset; Y \cap A = \emptyset) &= \exp[-\lambda(A)] \exp[-\nu(A)] \\
&= \exp[-(\lambda(A) + \nu(A))],
\end{aligned}
$$

using the independence of X and Y. The result follows from theorem 1.2. □

1.6 Finite point processes specified by a density

Although in most applications it is not realistic to assume that points are scattered randomly, Poisson processes are nevertheless useful as building blocks towards more complex models.

Example 1.11 Many aggregated data patterns in the life sciences may be thought of in an evolutionary manner as the offspring of some underlying 'parent' process. In such cases, a plausible model is a *Poisson cluster process* defined as follows. Suppose each point in a Poisson parent process gives birth to a finite point process of 'daughters', independently of other parents. Then the union of all daughters – provided it is locally finite – is a Poisson cluster process.

The special case where each daughter process consists of independent, identically distributed points is a *Neyman–Scott process* [151], originally introduced to describe the apparent aggregation of galaxies in cosmology [150]. For example, in the *Matérn cluster process* [129] illustrated in figure 1.10, each point in a homogeneous Poisson process X of intensity $\lambda > 0$ generates a Poisson number of daughters with mean ν, distributed uniformly in a ball of radius r around the parent point. By varying the distributions of the number of daughters and their spread, many other interesting processes can easily be obtained. For more details, see [16; 41; 131].

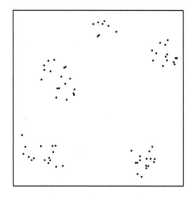

Fig. 1.10   Realisation of a Matérn process on $[0,1]^2$ with parent intensity $\lambda = 10$, mean number of daughters $\nu = 10$ per cluster and radius $r = 0.1$.

Example 1.12   The Poisson cluster process discussed in example 1.11 is a special case of a so-called *Boolean model*. In its simplest form, a Boolean model is defined as the union of balls of fixed radius $r > 0$ centered around the points of a homogeneous Poisson process X. Historically, this model has received a lot of attention because of its military interpretation [27; 59; 108; 129; 184; 185; 192; 205] as the devastation caused by randomly dropped bombs destroying everything within some given distance to the impact. For this reason the model used to be known as the *bombing model*, and was often expressed in terms of the Boolean function

$$q(x) = \begin{cases} 1 & \text{if } d(x, X) \leq r \\ 0 & \text{otherwise} \end{cases}$$

where $d(x, X)$ denotes the minimum distance between x and a point of X. Later, the Fontainebleau school around Matheron [130] coined the term Boolean model for any union of independent random compact sets (such as the offspring in example 1.11) generated by an underlying Poisson process. In this vein, figure 1.11 was obtained by placing a disc of random, exponentially distributed radius around each point in a realisation of a homogeneous Poisson process on $[0, 1]^2$ with intensity $\lambda = 1000$.

For more details, and references to recent work, see [130; 145; 195; 208].

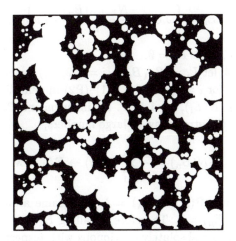

Fig. 1.11   Realisation of a Boolean model of discs on $[0, 1]^2$ with intensity $\lambda = 1000$ and exponential radii of mean $\frac{1}{75}$ digitised in a 512 by 512 image.

In the previous examples, hybrid models based on a Poisson process were presented. The remainder of this section is devoted to constructing point process models by means of their probability density (Radon-Nikodym derivative) with respect to a Poisson process. This approach is especially useful for finite processes. Indeed, as illustrated by lemma 1.1 below [208], absolute continuity for non-finite processes is a strong assumption. Moreover, given a collection of consistent probability densities on a sequence of sets increasing to $\mathbb{R}^d$ (say), a limit distribution does not always exist, and even if it does, is not necessarily unique [98; 131; 165].

Lemma 1.1   Let X_λ, X_μ be homogeneous Poisson processes on $\mathbb{R}^d$, defined on the same probability space $(\Omega, \mathcal{A}, P)$, with intensity λ and μ respectively. If $\lambda \neq \mu$, then the distribution of X_λ is not absolutely continuous with respect to that of X_μ.

Proof.   Consider the family $(B_n)_n$ of concentric closed balls centred at the origin with radii such that the volume of B_n equals n. For all $\nu > 0$, set

$$E_\nu = \left\{ \omega \in \Omega : \frac{N_{X_\nu(\omega)}(B_n)}{n} \to \nu \right\}.$$

Then, since $\lambda \neq \nu$, $E_\lambda \cap E_\nu = \emptyset$.

Setting $B_0 = \emptyset$, let $L_i := B_i \setminus B_{i-1}$ $(i = 1, 2, \dots)$ denote the increments. Then $L_i \cap L_j = \emptyset$ for $i \neq j$, and, by (P1)–(P2) in definition 1.5, under the distribution of X_λ, $N(L_i)$, $i = 1, 2, \dots$ are independent, Poisson distributed with mean λ. By the strong law of large numbers,

$$\frac{1}{n} N(B_n) = \frac{1}{n} \sum_{i=1}^{n} N(L_i) \to \mathbf{E} L_1 = \lambda$$

almost surely, hence $\mathbf{P}(X_\lambda \in E_\lambda) = 1$. The same arguments applied to X_μ yield $\mathbf{P}(X_\mu \in E_\mu) = 1$. Therefore, recalling E_λ and E_ν are disjoint, it follows that X_λ is not absolutely continuous with respect to X_ν.   □

In the remainder of this section, let $(\mathcal{X}, d)$ be a complete, separable metric space, and $\pi(\cdot)$ the distribution of a Poisson process on $\mathcal{X}$ with finite, non-atomic intensity measure $\nu(\cdot)$. Let $p : N^f \to [0, \infty)$ be a non-negative, measurable function on the family N^{f} of finite point configurations such that

$$\int_{N^{\mathrm{f}}} p(\mathbf{x})\, d\pi(\mathbf{x}) = 1. \tag{1.5}$$

Then $p(\cdot)$ is a probability density and defines a point process X on $\mathcal{X}$. Since the dominating Poisson process is finite and simple, so is X.

To interpret formula (1.5), write N^{f} as a union

$$N^{\mathrm{f}} = \cup_{n=0}^{\infty} N_n^{\mathrm{f}}$$

over the families N_n^{f} of configurations consisting of exactly n points. The ν-mass of N_n^{f} is $\nu(\mathcal{X})^n/n!$; the factor $n!$ is needed since $\mathcal{X}^n$ is ordered while N_n^{lf} is not. Thus, the mass of N^{f} equals

$$\sum_{n=0}^{n} \frac{\nu(\mathcal{X})^n}{n!} = e^{\nu(\mathcal{X})},$$

and, in order to obtain a probability distribution, a normalisation factor $e^{-\nu(\mathcal{X})}$ is called for.

From the above considerations, it follows that the distribution of the total number of points in a process X specified by density $p(\cdot)$ is given by

$$p_n = \frac{e^{-\nu(\mathcal{X})}}{n!} \int_{\mathcal{X}} \cdots \int_{\mathcal{X}} p(\{x_1,\ldots,x_n\})\, d\nu(x_1) \cdots d\nu(x_n), \tag{1.6}$$

$n \in \mathbb{N}_0$, and, conditional on the event $\{N(\mathcal{X}) = n\}$, the n random points have joint probability density

$$j_n(x_1,\ldots,x_n) = \frac{p(\{x_1,\ldots,x_n\})}{\int_{\mathcal{X}} \cdots \int_{\mathcal{X}} p(\{x_1,\ldots,x_n\})\, d\nu(x_1) \cdots d\nu(x_n)} \tag{1.7}$$

with respect to the n-fold product measure ν^n.

In the following examples, we shall derive densities for the Poisson and Matérn models on a compact set $\mathcal{X} \subseteq \mathbb{R}^d$.

Example 1.13 Let $\mathcal{X}$ be a compact subset of $\mathbb{R}^d$ with positive volume $\mu(\mathcal{X})$, and consider a Poisson process X of rate λ on $\mathcal{X}$. To find its density at $\mathbf{x} = \{x_1,\ldots,x_n\}$ with respect to a unit rate Poisson process, heuristically one has to consider the ratio

$$\frac{q_\lambda(\{x_1, \cdots, x_n\})}{q_1(\{x_1, \cdots, x_n\})}$$

where $q_\lambda(\{x_1, \cdots, x_n\}) \, dx_1 \ldots dx_n$ denotes the 'probability' that X has exactly n points, one each at the infinitesimal regions $dx_1, \ldots, dx_n$. Now, for intensity parameter λ, the probability that $N(\mathcal{X})$ equals n is $p_n = e^{-\lambda\mu(\mathcal{X})} \frac{(\lambda\mu(\mathcal{X}))^n}{n!}$ while the conditional probability of the n points falling in $dx_1, \ldots, dx_n$ is $n! \frac{dx_1 \ldots dx_n}{\mu(\mathcal{X})^n}$. Hence our candidate density is

$$p(\mathbf{x}) = e^{(1-\lambda)\mu(\mathcal{X})} \lambda^n. \tag{1.8}$$

In order to verify that (1.8) indeed defines a Poisson process, note that by (1.6)

$$\begin{aligned}
\mathbf{P}(N(\mathcal{X}) = n) &= \frac{e^{-\mu(\mathcal{X})}}{n!} \int_{\mathcal{X}} \cdots \int_{\mathcal{X}} e^{(1-\lambda)\mu(\mathcal{X})} \lambda^n \, dx_1 \cdots dx_n \\
&= e^{-\lambda\mu(\mathcal{X})} \frac{(\lambda\mu(\mathcal{X}))^n}{n!}.
\end{aligned}$$

Furthermore, using (1.7),

$$j_n(x_1, \ldots, x_n) = \mu(\mathcal{X})^{-n},$$

a density of the uniform distribution on $\mathcal{X}$. Hence $p(\cdot)$ specifies a Poisson process of rate λ.

Example 1.14   As in the previous example, let $\mathcal{X}$ be a compact subset of $\mathbb{R}^d$ with positive volume $\mu(\mathcal{X})$, and consider the Matérn model introduced in example 1.11. Note that the offspring of a parent process in $\mathcal{X}$ form a subset of $\mathcal{Y} = \mathcal{X}_{(r)}$, the r-envelope of $\mathcal{X}$. Moreover, conditionally on a parent configuration $\mathbf{x} = \{x_1, \ldots, x_n\}$, the daughters form an inhomogeneous Poisson process on $\mathcal{Y}$ with intensity function

$$\lambda(y \mid \mathbf{x}) = \frac{\nu}{\pi r^2} \sum_{i=1}^{n} \mathbb{I}\{y \in B(x_i, r)\},$$

where $B(x_i, r)$ is a closed ball of radius r centred at x_i, and ν the mean number of offspring per cluster. By arguments similar to those used in example 1.13, the offspring is absolutely continous with respect to a unit rate

Poisson process on $\mathcal{Y}$, with conditional density $p(\mathbf{y}|\mathbf{x})$ at $\mathbf{y} = \{y_1, \ldots, y_m\}$ given by

$$\exp\left[\int_{\mathcal{Y}}\left(1 - \frac{\nu}{\pi r^2}\sum_{i=1}^{n}\mathbb{I}\{t \in B(x_i, r)\}\right)dt\right]\prod_{j=1}^{m}\left(\frac{\nu}{\pi r^2}\sum_{i=1}^{n}\mathbb{I}\{y_j \in B(x_i, r)\}\right).$$

The unconditional density can be found by taking the expectation over $\mathbf{x}$:

$$p(\mathbf{y}) = \sum_{n=0}^{\infty}\frac{e^{-\lambda\mu(\mathcal{X})}}{n!}\lambda^n\int_{\mathcal{X}}\cdots\int_{\mathcal{X}}p(\mathbf{y} \mid \{x_1, \ldots, x_n\})\,dx_1\cdots dx_n.$$

More general cluster processes are considered in [10] and [117; 121]; probability densities for Boolean models are discussed in [118].

Not all finite point processes are absolutely continuous with respect to a given Poisson model. For instance, the sphere having volume zero, the model of example 1.7 is not dominated by a unit rate Poisson process on $[-1, 1]^2$. Other examples include randomly translated grids, or planar cluster processes where the offspring of a point at x is $\{x + a, x - a\}$ for some fixed a.

1.7 Campbell and moment measures

The mean and higher order moments are important entities in the theory of random vectors. For point processes, it is not so easy to give a definition of 'mean pattern'. Nevertheless, by definition 1.1, $N(A)$ is a random variable for each bounded Borel set A, and the moments of $N(A)$ provide a useful collection of summary statistics. More generally, we can restrict attention to configurations with specified properties by considering the moments of $N_X(A)\,\mathbb{I}\{X \in F\}$. Below, we will use these ideas to define the *moment* and *Campbell measure* [29; 41; 82; 91; 98; 113; 131; 134; 135] of a point process.

Definition 1.6 Let X be a point process on a complete, separable metric space $(\mathcal{X}, d)$. Define

$$M(A) = \mathbf{E}N(A)$$

and

$$C(A \times F) = \mathbf{E}\left[N(A)\, \mathbb{I}\{X \in F\}\right]$$

for any bounded Borel set $A \subseteq \mathcal{X}$ and $F \in \mathcal{N}^{\text{lf}}$.

The functions $M(\cdot)$ and $C(\cdot \times \cdot)$ are not necessarily finite as demonstrated by the following example.

Example 1.15 Let X be a finite point process on a compact subset $\mathcal{X} \subseteq \mathbb{R}^d$ defined by

$$p_n = \begin{cases} \frac{1}{n(n-1)} & n \geq 2 \\ 0 & n = 0, 1 \end{cases}$$

and any family of permutation invariant probability densities $j_n(\cdot, \ldots, \cdot)$ (cf. section 1.4). Then $M(\mathcal{X}) = \sum_{n=2}^{\infty} n\, p_n = \sum_{n=1}^{\infty} \frac{1}{n} = \infty$.

The next lemma gives conditions under which the set functions $M(\cdot)$ and $C(\cdot \times \cdot)$ can be extended to measures on the Borel σ-algebra $\mathcal{B}(\mathcal{X})$ of $\mathcal{X}$ and the product σ-algebra $\mathcal{B}(\mathcal{X}) \times \mathcal{N}^{\text{lf}}$ respectively [41; 131].

Lemma 1.2   *If the function $M(\cdot)$ introduced in definition 1.6 is finite on bounded Borel sets, then $M(\cdot)$ can be extended uniquely to a σ-finite measure on the Borel sets of $\mathcal{X}$, the first-order moment measure.*

The function $C(\cdot \times \cdot)$ (definition 1.6) can be extended uniquely to a σ-finite measure on the product σ-algebra of $\mathcal{B}(\mathcal{X})$ and $\mathcal{N}^{\text{lf}}$, the first-order Campbell measure.

Proof.   To show that $M(\cdot)$ is a measure, let the bounded Borel set A be a countably infinite union of disjoint bounded Borel sets $A_1, A_2, \ldots$. Then

$$M(A) = \mathbf{E}\left[\sum_{i=1}^{\infty} N(A_i)\right] = \sum_{i=1}^{\infty} \mathbf{E}N(A_i) = \sum_{i=1}^{\infty} M(A_i)$$

by the monotone convergence theorem [79]. Thus $M(\cdot)$ is countably additive and finite (by assumption) on the ring of bounded Borel sets. Hence, by the measure extension theorem [79], $M(\cdot)$ can be extended uniquely to a σ-finite measure on the Borel σ-algebra.

Regarding the Campbell measure, let A be a bounded Borel set and F an element of $\mathcal{N}^{\text{lf}}$. Suppose that $A \times F$ can be written as a countably infinite union of disjoint product sets $A_1 \times F_1, A_2 \times F_2, \ldots$, where A_i is bounded Borel and $F_i \in \mathcal{N}^{\text{lf}}$. Then

$$C(A \times F) = \mathbf{E}\left[\sum_{i=1}^{\infty} N(A_i)\, \mathbb{I}\{X \in F_i\}\right] = \sum_{i=1}^{\infty} C(A_i \times F_i)$$

by the monotone convergence theorem. Thus $C(\cdot \times \cdot)$ is a measure on the semi-ring of product sets. To show that $C(\cdot \times \cdot)$ is σ-finite, let A_m, $m = 1, 2, \ldots$ be a countable covering of $\mathcal{X}$ consisting of bounded Borel sets. Write, for $m, n \in \mathbb{N}$,

$$F_{mn} = \left\{\omega \in \Omega : N_{X(\omega)}(A_m) \le n\right\}.$$

Then

$$C(A_m \times F_{mn}) = \mathbf{E}\left[N(A_m)\, \mathbb{I}\{X \in F_{mn}\}\right] \le n\, \mathbf{P}(F_{mn}) \le n < \infty.$$

To see that $\mathcal{X} \times \mathcal{N}^{\text{lf}}$ is covered by $(A_m \times F_{mn})_{m,n}$, pick any $(x, \mathbf{x}) \in \mathcal{X} \times \mathcal{N}^{\text{lf}}$. Since the A_m cover $\mathcal{X}$, there exists an A_m such that $x \in A_m$. Moreover, as $\mathbf{x}$ is locally finite, $N(A_m) < \infty$, therefore $N(A_m) \le n$ for large enough n and $(x, \mathbf{x}) \in A_m \times F_{mn}$. Hence $C(\cdot \times \cdot)$ is σ-finite on the product sets. Finally, again applying the measure extension theorem, $C(\cdot \times \cdot)$ can be extended uniquely to a σ-finite measure $C(\cdot)$ on $\mathcal{B}(\mathcal{X}) \times \mathcal{N}^{\text{lf}}$, the σ-algebra generated by the product sets. $\qquad\square$

The first order moment measure can be expressed in terms of the Campbell measure as follows:

$$C(A \times \mathcal{N}^{\text{lf}}) = \mathbf{E}N(A) = M(A)$$

for any Borel set A. Therefore $C(\cdot)$ is finite if and only if $M(\cdot)$ is finite, that is $\mathbf{E}N(\mathcal{X}) < \infty$.

Higher order Campbell measures are defined in a similar fashion [98], for instance

$$C^{(2)}(A \times B \times F) = \mathbf{E}\left[N(A)\, N(B)\, \mathbb{I}\{X \in F\}\right]$$

$(A, B$ bounded Borel sets, $F \in \mathcal{N}^{lf})$ which is related to the second order moment measure by

$$C^{(2)}(A \times B \times \mathcal{N}^{lf}) = \mathbf{E}\left[N(A)N(B)\right].$$

Example 1.16   The first order moment measure of a binomial point process (see example 1.6) is

$$M(A) = n\,\mathbf{P}(X_i \in A) = n\,\frac{\mu(A)}{\mu(\mathcal{X})},$$

for any Borel set $A \subseteq \mathcal{X}$.

Example 1.17   The first order moment measure of a homogeneous Poisson process on a compact set $\mathcal{X} \subseteq \mathbb{R}^d$ of positive volume is

$$M(A) = \lambda\mu(A),$$

where A is a Borel subset of $\mathcal{X}$, and λ the intensity of the process.   For inhomogeneous Poisson processes, the moment measure is identical to the intensity measure.

Example 1.18   Further to example 1.17, the Campbell measure of a homogeneous Poisson process X is defined by

$$C(A \times F) = \lambda \int_A \int_F dP_x(\mathbf{x})dx$$

for bounded Borel sets $A \subseteq \mathcal{X}$ and $F \in \mathcal{N}^{lf}$. Here $P_x(\cdot)$ is the distribution of X with an additional (deterministic) point at x.

The above expression can be derived heuristically by noting that $C(A \times F) = \mathbf{E}[N(A)\,\mathbb{I}\{X \in F\}]$ by definition. Now for $x \in A$, the probability of a point in the infinitesimal region dx is $\lambda\,dx$; the probability of more than one point may be ignored. Moreover, given there is a point in dx, the remainder of the process is Poisson by the spatial randomness property. Hence the conditional probability that X falls in the event F is

$$\int_F dP_x(\mathbf{x}).$$

Summing over infinitesimal regions yields the result. A more rigorous derivation will be given in section 1.8.

Campbell measures are important tools when working with point processes. Suppose a measurement $g(x, \mathbf{x})$ is taken at each point x in a realisation $\mathbf{x}$ of X. The function $g(x, \mathbf{x})$ may depend on other points of the pattern $\mathbf{x}$. For instance in example 1.19, $g(x, \mathbf{x})$ is the distance to the nearest other point in $\mathbf{x}$. Situations where $g(\cdot, \cdot)$ is a function of location x solely arise in forestry, where $g(x)$ may denote the stem diameter of a tree at x, or in environmental statistics where $g(x)$ could be measuring covariates such as the concentration of a pollutant. By theorem 1.5, the expected total $\mathbf{E}\left[\sum_{x \in X} g(x, X)\right]$ is simply the integral of $g(\cdot)$ with respect to the Campbell measure [41; 131; 168; 208]. From a theoretical point of view, the integral representation of the Campbell measure implicit in (1.9) will be used to define the conditional intensity of a point process (see section 1.8).

Theorem 1.5 (Campbell–Mecke formula)
Let $g : \mathcal{X} \times N^{\mathrm{lf}} \to \mathbb{R}$ be a measurable function that is either non-negative or integrable with respect to the Campbell measure. Then

$$\mathbf{E}\left[\sum_{x \in X} g(x, X)\right] = \int_{\mathcal{X}} \int_{N^{\mathrm{lf}}} g(x, X) \, dC(x, X). \qquad (1.9)$$

If the function g does not depend on the point process X, the Campbell–Mecke formula reduces to

$$\mathbf{E}\left[\sum_{x \in X} g(x)\right] = \int_{\mathcal{X}} g(x) \, dM(x) \qquad (1.10)$$

provided the first order moment measure $M(\cdot)$ exists and is finite on bounded Borel sets. In particular, for a homogeneous Poisson process of intensity $\lambda > 0$,

$$\mathbf{E}\left[\sum_{x \in X} g(x)\right] = \lambda \int_{\mathcal{X}} g(x) \, dx$$

is proportional to the integral of $g(\cdot)$, see [29; 131].

Proof. Firstly, let $g(\cdot, \cdot)$ be an indicator function $g(x, \mathbf{x}) = \mathbb{I}\{(x, \mathbf{x}) \in A \times F\}$ for some bounded Borel set A and some $F \in \mathcal{N}^{lf}$. Then

$$\mathbf{E}\left[\sum_{x \in X} g(x, X)\right] = \mathbf{E}\left[N(A)\,\mathbb{I}\{X \in F\}\right] = C(A \times F) =$$

$$\int_{\mathcal{X}}\int_{\mathcal{N}^{lf}} \mathbb{I}\{(x, \mathbf{x}) \in A \times F\}\,dC(x, \mathbf{x}) = \int_{\mathcal{X}}\int_{\mathcal{N}^{lf}} g(x, \mathbf{x})\,dC(x, \mathbf{x}).$$

Using the linearity of the expectation and integral, the Campbell–Mecke formula holds for step functions. Since the product sets generate $\mathcal{B}(\mathcal{X}) \times \mathcal{N}^{lf}$, by the monotone convergence theorem, (1.9) is true for $g(x, \mathbf{x}) = \mathbb{I}\{(x, \mathbf{x}) \in B\}$ for any $B \in \mathcal{B}(\mathcal{X}) \times \mathcal{N}^{lf}$. Consequently, (1.9) holds for functions of the form $\sum_{i=1}^{n} \beta_i\,\mathbb{I}\{(x, \mathbf{x}) \in B_i\}$ and, by taking limits, for any non-negative or integrable function $g(\cdot, \cdot)$. $\qquad\square$

Example 1.19 The Campbell–Mecke formula provides an alternative way to calculate the mean number of r-close pairs in a homogeneous Poisson process (cf. example 1.8). Thus let X be a Poisson point process with intensity $\lambda > 0$ on $[0, 1]^2$. The expected number of r-close pairs can be written as

$$\mathbf{E}\left[\frac{1}{2}\sum_{x \in X} \#\{y \in X : 0 < \|x - y\| \le r\}\right] = \mathbf{E}\left[\sum_{x \in X} g(x, X)\right]$$

for $g(x, X) = \frac{1}{2}\#\{y \in X : 0 < \|x - y\| \le r\}$. By the Campbell–Mecke formula and example 1.18,

$$\mathbf{E}\left[\sum_{x \in X} g(x, X)\right] = \int_{[0,1]^2}\int_{\mathcal{N}^{lf}} g(x, \mathbf{x})\,dC(x, \mathbf{x})$$

$$= \frac{\lambda}{2}\int_{[0,1]^2}\left[\int_{\mathcal{N}^{lf}} \#\{y \in \mathbf{x} : 0 < \|x - y\| \le r\}\,dP_x(\mathbf{x})\right]dx. \qquad (1.11)$$

The inner integrand equals

$$\lambda\int_{[0,1]^2} \mathbb{I}\{\|y - x\| \le r\}\,dy,$$

hence (1.11) reduces to

$$\frac{1}{2}\lambda^2 \int_{[0,1]^2} \int_{[0,1]^2} \mathbb{I}\{\|y - x\| \le r\}\, dx\, dy.$$

1.8 Interior and exterior conditioning

In this section we describe how to define rigorously the conditional distribution of a point process X at a particular point $x \in \mathcal{X}$ given its configuration on $\mathcal{X} \setminus \{x\}$ (exterior conditioning) and the conditional distribution of the process given there is a point at $x \in \mathcal{X}$ (interior conditioning). The two concepts are dual; exterior conditioning is formalised by the *Papangelou conditional intensity* [161], interior conditioning by the *Palm distribution* [160].

1.8.1 *A review of Palm theory*

From now on, assume the first order moment measure $M(\cdot)$ exists and is σ-finite (cf. lemma 1.2). Then for each $F \in \mathcal{N}^{\text{lf}}$, the marginal Campbell measure with second argument fixed at F is absolutely continuous with respect to the moment measure, that is

$$C(\cdot \times F) << M(\cdot).$$

Hence, a Radon-Nikodym derivative can be introduced as a non-negative Borel measurable function $P_\cdot(F) : \mathcal{X} \to \mathbb{R}$ such that for each Borel set $A \in \mathcal{B}(\mathcal{X})$

$$C(A \times F) = \int_A P_x(F)\, dM(x).$$

Note that $P_\cdot(F)$ is defined uniquely up to an M-null set. Moreover, it is possible to find a version $P_x(F)$ such that for fixed $x \in \mathcal{X}$ and variable F, $P_x(\cdot)$ is a probability distribution, while for variable x and fixed F, $P_\cdot(F)$ is a Borel measurable function. The probability distributions $P_x(\cdot)$ are the *Palm distributions* of X at $x \in \mathcal{X}$ [41; 98; 99; 131]. They can be interpreted as the conditional distributions of X given $N(\{x\}) > 0$.

The Campbell–Mecke formula (1.9) can be rewritten in terms of the Palm distribution as

$$\mathbf{E}\left[\sum_{x \in X} g(x, X)\right] = \int_{\mathcal{X}} \int_{N^{\mathrm{lf}}} g(x, \mathbf{x})\, dP_x(\mathbf{x})\, dM(x). \qquad (1.12)$$

Example 1.20 In example 1.18 we argued intuitively that the Palm distribution $P_x(\cdot)$ of a homogeneous Poisson process with distribution $P(\cdot)$ is the convolution $P * \delta_x(\cdot)$ of $P(\ )$ with an additional deterministic point at x. In fact, this form of the Palm distribution characterises the family of Poisson processes [41; 131; 208].

More generally, let $P^\nu(\cdot)$ be the distribution of a Poisson process with locally finite, non-atomic intensity measure $\nu(\cdot)$. Since both the Palm distribution $P_x^\nu(\cdot)$ and $P^\nu * \delta_x(\cdot)$ are simple, by theorem 1.2 it is sufficient to prove that their void probabilities coincide. To do so, pick a bounded Borel set A and write $v_x(\cdot)$ for the void probabilities corresponding to $P^\nu * \delta_x(\cdot)$, and $v(\cdot)$ for those of the Poisson process. Then for any bounded Borel set $B \subseteq \mathcal{X}$, by definition

$$C(A \times \{N(B) = 0\}) = \int_A P_x^\nu(N(B) = 0)\, d\nu(x).$$

On the other hand

$$\begin{aligned}
C(A \times \{N(B) = 0\}) &= C(A \setminus B \times \{N(B) = 0\}) = v(B)\,\nu(A \setminus B) \\
&= \int_{A \setminus B} v(B)\, d\nu(x) = \int_A v_x(B)\, d\nu(x),
\end{aligned}$$

using the fact that for a Poisson process the random variables $N(A \setminus B)$ and $N(B)$ are independent. Since A was chosen arbitrarily, $P_x^\nu(N(B) = 0) = (P^\nu * \delta_x)(N(B) = 0)$ for ν-almost all x.

Many useful point process statistics are defined in terms of the Palm distribution. Let X be a *stationary* point process on $\mathbb{R}^d$ (that is, its distribution is translation invariant and hence does not change if all points of X are translated over some vector $y \in \mathbb{R}^d$). Then the *nearest-neighbour distance distribution function* [46; 219] of X is defined by

$$G(r) = \mathbf{P}_y(d(y, X \setminus \{y\}) \le r), \qquad r \ge 0, \qquad (1.13)$$

the probability that X places at least one point within distance r of y. The translation invariance of the distribution of X is inherited by the Palm distribution, implying that $G(r)$ is well-defined and does not depend on the choice of y. Replacing the Palm distribution in (1.13) by the distribution of X, we obtain the *empty space function* [46]

$$F(r) = \mathbf{P}(d(y, X) \leq r), \qquad r \geq 0.$$

Thus, $F(\cdot)$ is the distribution function of the distance from an arbitrary point $y \in \mathbb{R}^d$ to the nearest point of the process. Again, the definition of $F(r)$ does not depend on the choice of y. The ratio

$$J(r) = \frac{1 - G(r)}{1 - F(r)}, \qquad (1.14)$$

defined for all $r \geq 0$ such that $F(r) < 1$, compares nearest–neighbour to empty space distances [122]. It takes values less than 1 whenever the empty spaces tend to be larger than the distance between nearest neighbour pairs, thus indicating clustering, whereas values exceeding 1 suggest a more regular pattern.

Example 1.21 For a Poisson process of intensity λ on $\mathbb{R}^d$

$$1 - F(r) = \exp\left[-\lambda\mu(B(y, r))\right]$$

the probability that the closed ball $B(y, r)$ of radius r around y contains no points of X. By example 1.20, $F(r) = G(r)$ for all $r \geq 0$, hence $J \equiv 1$.

It should be noted that the list of summary statistics discussed here is by no means complete; a variety of derived as well as higher order statistics have been suggested, see for instance [46; 123; 173; 174; 194; 200; 208; 209; 219]. For multivariate point processes, cross versions of the statistics described above can be used [3; 46; 123; 125; 60; 85; 93; 209; 214]. For instance, the cross nearest–neighbour distance distribution function measures distances from a point marked as i to the nearest point labelled j.

Example 1.22 In Figure 1.12 (dashed lines), we have plotted estimates [5] of the J–statistic for the point configurations in Figures 1.9 and 1.10. For

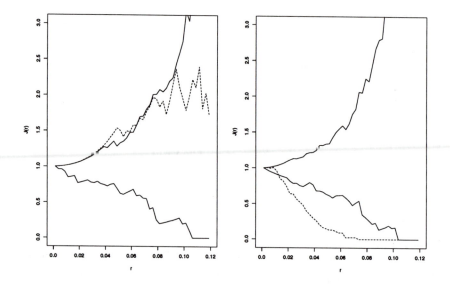

Fig. 1.12 Estimated J functions (dashed line) for the patterns in Figures 1.9 (left) and 1.10 (right). The solid lines are upper and lower envelopes based on 99 independent simulations of a binomial process with the same number of points as the data pattern.

the thinned pattern, as expected, the estimated J–function takes values greater than 1, indicating an ordered pattern. For the Matérn process on the other hand, $J(r) < 1$ in accordance with the clustered nature of the model.

Often in the literature, graphs of estimated summary statistics are embellished by upper and lower envelopes obtained by simulating a number of independent realisations of a null hypothesis model (for instance a binomial process with the same number of points as the data), computing the statistic at hand for each realisation, and recording the maximum and minimum value. As well as giving an indication of the variability, such simulation envelopes are useful for testing purposes. For instance, a Monte Carlo test [21; 45; 46; 112; 173; 174; 175; 176; 179] for a binomial null hypothesis can be based on

$$U_1 = \int_0^S (\hat{J}_1(r) - 1)^2 \, dr$$

where $\hat{J}_1(r)$ is the estimated J–function (1.14) of the data pattern and S a threshold value. The idea is to compare the value of U_1 for the data pattern to values $U_2, \dots, U_n$ obtained from $n - 1$ independent simulations of the binomial process. Rejecting the null hypothesis whenever the rank of U_1 is at least k yields an exact test at level k/n, since under the null hypothesis each rank is equally likely.

Functions such as $F(\cdot), G(\cdot)$ and $J(\cdot)$ provide valuable information on the process and have been applied successfully. Nevertheless, it is worth keeping in mind that no low–dimensional statistic fully characterises the distribution of the process [12; 18].

1.8.2 *A review of conditional intensities*

Turning attention to the dual case of exterior conditioning, the *Papangelou conditional intensity* $\lambda(\cdot; \cdot)$ associated with a simple point process X can be interpreted as

$$\lambda(x; \mathbf{x}) \, dx = \mathbf{P}(N(dx) = 1 \mid X \cap (dx)^c = \mathbf{x} \cap (dx)^c),$$

the infinitesimal probability of finding a point in a region dx around $x \in \mathcal{X}$ given that the point process agrees with the configuration $\mathbf{x}$ outside of dx. In order to give a more precise definition, we need the concept of *reduced Campbell measures*[41; 98; 99; 131; 208].

Definition 1.7 Let X be a simple point process on the complete, separable metric space $(\mathcal{X}, d)$. Define

$$C^!(A \times F) = \mathbf{E} \left[\sum_{x \in X \cap A} \mathbb{I}\{X \setminus \{x\} \in F\} \right]$$

for any bounded Borel set $A \subseteq \mathcal{X}$ and $F \in \mathcal{N}^{\text{lf}}$.

Arguing as in section 1.7, the function $C^!(\cdot \times \cdot)$ may be extended uniquely to a σ-finite measure on the product σ-algebra $\mathcal{B}(\mathcal{X}) \times \mathcal{N}^{\text{lf}}$. The analogue of the Campbell–Mecke formula reads

$$\mathbf{E} \left[\sum_{x \in X} g(x, X \setminus \{x\}) \right] = \int_{\mathcal{X}} \int_{N^{\text{lf}}} g(x, \mathbf{x}) \, dC^!(x, \mathbf{x}). \tag{1.15}$$

for any measurable function $g(\cdot, \cdot)$ that is either non-negative or integrable.

Assuming the first order moment measure $M(\cdot)$ of X exists and is σ-finite (cf. lemma 1.2), we can apply Radon-Nikodym theory to write

$$C^!(A \times F) = \int_A P_x^!(F)\, dM(x)$$

for any Borel set $A \subseteq \mathcal{X}$ and $F \in \mathcal{N}^{!f}$. The function $P_x^!(F)$ is defined uniquely up to an M-null set. Moreover, it is possible to find a version such that for fixed $x \in \mathcal{X}$, $P_x^!(\cdot)$ is a probability distribution, the *reduced Palm distribution* at x. It can be interpreted as the conditional distribution of the remainder $X \setminus \{x\}$ of the point process, given $N(\{x\}) > 0$.

Example 1.23   From example 1.20 it follows immediately that the reduced Palm distribution of a homogeneous Poisson process with intensity λ on $\mathbb{R}^d$ is the same as its distribution $P(\cdot)$. Hence for any bounded Borel set $A \subseteq \mathcal{X}$, and any $F \in \mathcal{N}^{!f}$,

$$C^!(A \times F) = \lambda \int_A P_x^!(F)\, dx = \lambda\, P(F)\, \mu(A)$$

where $\mu(A)$ denotes the volume of A.

Make the further assumption that for any fixed bounded Borel set $A \in \mathcal{B}(\mathcal{X})$, $C^!(A \times \cdot)$ is absolutely continuous with respect to the distribution $P(\cdot)$ of X. Then

$$C^!(A \times F) = \int_F \Lambda(A; \mathbf{x})\, dP(\mathbf{x})$$

for some measurable function $\Lambda(A; \cdot)$, specified uniquely up to a P-null set. Moreover, one can find a version such that for fixed $\mathbf{x}$, $\Lambda(\cdot; \mathbf{x})$ is a locally finite Borel measure [41; 98; 99; 131], the first order *Papangelou kernel*.

If $\Lambda(\cdot; \mathbf{x})$ admits a density $\lambda(\cdot; \mathbf{x})$ with respect to a reference measure $\nu(\cdot)$ on $\mathcal{X}$, (1.15) factorises as

$$\mathbf{E}\left[\int_{\mathcal{X}} g(x, X)\, \lambda(x; X)\, d\nu(x)\right] ; \tag{1.16}$$

the function $\lambda(\cdot; \cdot)$ is called the *Papangelou conditional intensity*.

The case where the distribution of X is dominated by a Poisson process is especially important [67; 68; 110; 132; 152; 174; 179].

Theorem 1.6 *Let X be a finite point process specified by a density $p(\mathbf{x})$ with respect to a Poisson process with non-atomic finite intensity measure $\nu(\cdot)$. Then X has Papangelou conditional intensity*

$$\lambda(u; \mathbf{x}) = \frac{p(\mathbf{x} \cup \{u\})}{p(\mathbf{x})} \tag{1.17}$$

for $u \notin \mathbf{x} \in N^{\mathrm{f}}$.

Heuristically, (1.17) can be derived by noting that the infinitesimal probability of the configuration $\mathbf{x} \cup \{u\}$ is $e^{-\nu(\mathcal{X})} p(\mathbf{x} \cup \{u\}) \, dx_1 \cdots dx_n \, du$. Similarly, the infinitesimal probability of $\mathbf{x}$ on the complement of du equals $e^{-\nu(\mathcal{X})} \, dx_1 \cdots dx_n (p(\mathbf{x} \cup \{u\}) \, du + p(\mathbf{x}))$; the term $p(\mathbf{x} \cup \{u\}) \, du$ is negligable compared to $p(\mathbf{x})$, so the conditional intensity of a point at u given the configuration $\mathbf{x} = \{x_1, \ldots, x_n\}$ elsewhere must be (1.17).

Proof. We will use the integral characterisation (1.15)–(1.16). Thus, let $g : \mathcal{X} \times N^{\mathrm{lf}} \to \mathbb{R}$ be a non-negative or integrable function. Substitution of (1.17) in (1.16) and the fact that the distribution of X is given by means of a density $p(\cdot)$ yield

$$\mathbf{E}\left[\int_{\mathcal{X}} g(u, X)\, \lambda(u; X)\, d\nu(u)\right] =$$

$$\sum_{n=0}^{\infty} \frac{e^{-\nu(\mathcal{X})}}{n!} \int \cdots \int_{\mathcal{X}^{n+1}} g(u, \{x_1, \ldots, x_n\}) \frac{p(\{x_1, \ldots, x_n\} \cup \{u\})}{p(\{x_1, \ldots, x_n\})}$$

$$p(\{x_1, \ldots, x_n\})\, d\nu(x_1) \cdots d\nu(x_n)\, d\nu(u) =$$

$$\sum_{n=0}^{\infty} \frac{e^{-\nu(\mathcal{X})}}{n!} \frac{1}{n+1} \int \cdots \int_{\mathcal{X}^{n+1}} \sum_{i=1}^{n+1} g(x_i, \{x_1, \ldots, x_{n+1}\} \setminus \{x_i\})$$

$$p(\{x_1, \ldots, x_{n+1}\})\, d\nu(x_1) \ldots d\nu(x_{n+1}) =$$

$$\sum_{n=1}^{\infty} \frac{e^{-\nu(\mathcal{X})}}{n!} \int \cdots \int_{\mathcal{X}^{n}} \sum_{i=1}^{n} g(x_i, \{x_1, \ldots, x_n\} \setminus \{x_i\})$$

$$p(\{x_1, \ldots, x_n\})\, d\nu(x_1) \cdots d\nu(x_n). \tag{1.18}$$

Since (1.18) equals the expectation in the left hand side of (1.15), and $g(\cdot, \cdot)$ was chosen arbitrarily, we conclude that $\lambda(\cdot; \cdot)$ satisfies the integral representation and the proof is complete. $\square$

The Papangelou conditional intensity is very useful to describe the local interactions in a point pattern, and as such leads naturally to the notion of a *Markov point process* in chapter 2. On a more elementary level, if $\lambda(x; \mathbf{x}) \equiv \lambda(x; \emptyset)$ for all patterns $\mathbf{x}$ satisfying $\mathbf{x} \cap B(x, R) = \emptyset$, we say that the process has 'interactions of range R at x'. In other words, points further than R away from x do not contribute to the conditional intensity at x.

Example 1.24   Suppose that X is a stationary point process on $\mathbb{R}^d$ whose Papangelou conditional intensity $\lambda(\cdot; \cdot)$ exists. If X has interactions of range R, the J–function (1.14) is constant beyond R.

To see this [122], let $F = \{X \cap B(0, r) = \emptyset\}$, so that $1 - F(r) = \mathbf{P}(F)$. Furthermore, writing $\mathbf{E}_0^!$ for the expectation with respect to the reduced Palm distribution at the origin, $1 - G(r) = \mathbf{E}_0^!(F)$. Let A be a bounded Borel set, and define

$$g(x, \mathbf{x}) = \frac{\mathbb{I}\{x \in A; \mathbf{x} \in F\}}{\lambda(0; \mathbf{x})}, \qquad x \in \mathcal{X}, \mathbf{x} \in N^{\text{lf}}.$$

Clearly, $g(\cdot, \cdot)$ is non-negative and measurable. Since X is stationary, the function $M(\cdot)$ of definition 1.6 is translation invariant, hence the first order moment measure exists and is proportional to Lebesgue measure $\mu(\cdot)$, i.e. $M(B) = \lambda\mu(B)$ for every Borel set B and some $\lambda \in \mathbb{R}^+$. Thus, (1.15)– (1.16) reduce to [152] $\lambda\mu(A) \mathbf{E}_0^! [\mathbb{I}\{X \in F\}/\lambda(0; X)] = \mu(A) \mathbf{P}(F)$. Since by assumption $\lambda(0; X) \equiv \lambda(0; \emptyset)$ on F, $\mathbf{E}_0^! [\mathbb{I}\{X \in F\}/\lambda(0; X)] = (1 - G(r))/\lambda(0; \emptyset)$, we obtain

$$J(r) = \frac{\lambda(0; \emptyset)}{\lambda} \qquad \text{for all } r \geq R.$$

Chapter 2

Markov Point Processes

2.1 Ripley–Kelly Markov point processes

Most of the models encountered in chapter 1 exhibit some kind of independence between the points. Indeed, in a binomial process all points are independent. For Poisson processes, the point counts of disjoint regions are independent, and a conditional independence property (theorem 1.3) holds. Both the Poisson cluster process introduced in example 1.11 and the Boolean model (example 1.12) are based on an underlying Poisson process of parents or germs.

The present chapter focuses on *Markov spatial point processes* [11; 109; 165; 181; 176; 179; 180; 191; 217]. This class of models is especially designed to take into account inter-point interactions, but includes the Poisson process and certain Poisson cluster models (see chapter 4) as well.

As a motivating example, imagine a molecular biologist is keen to model a pattern of cells observed through a microscope. Let us assume that the cells are all of the same type, and, to a good approximation, circular with some fixed radius $R > 0$. In these circumstances, the pattern can conveniently be summarised by listing the locations of the nuclei. It is obvious that no nucleus can be closer than $2R$ to another nucleus, hence, in the absence of any further interaction, a natural choice is the hard core density

$$p((\{x_1, \ldots, x_n\})) = \alpha \beta^n \; \mathbb{I}\{\|x_i - x_j\| > 2R, i \neq j\}. \tag{2.1}$$

Here $\beta > 0$ is an intensity parameter; the constant $\alpha > 0$ ensures $p(\cdot)$ integrates to 1. For each $n \in \mathbb{N}$, the right hand side is a Borel measurable symmetric function on $\mathbb{R}^n$, hence $p(\cdot)$ is measurable (cf. the discussion in the beginning of section 1.4). A typical realisation for $\beta = 100$ and $2R = 0.05$ is displayed in Figure 2.1.

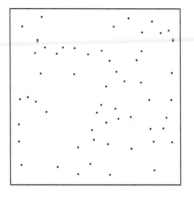

Fig. 2.1   Realisation of a hard core model in the unit square with $\beta = 100$ and hard core distance $2R = 0.05$.

Model (2.1) is defined in terms of empty overlap between idealised cells. Hence, to compute the Papangelou conditional intensity $\lambda(u; \mathbf{x})$ (cf. section 1.8), one needs to check whether a ball of radius $2R$ around u contains any other member of the set $\mathbf{x}$. Thus, the conditional behaviour at u depends *locally* on $\mathbf{x}$, a property reminiscent of Markov chains in time [75]. More specifically, for a discrete Markov chain $X_0, \ldots X_N$,

$$P(X_i = x_i \mid X_0 = x_0; \cdots ; X_{i-1} = x_{i-1}) = P(X_i = x_i \mid X_{i-1} = x_{i-1}),$$
$$(2.2)$$

$i = 1, \ldots, N$, the well-known conditional independence of the 'future' X_i and the 'past' $X_0, \ldots, X_{i-2}$ given the 'present' X_{i-1}. By the product rule, the joint distribution

$$P(X_0 = x_0; \cdots ; X_N = x_N) = P(X_0 = x_0) \prod_{i=1}^{N} p(x_{i-1}, x_i) \qquad (2.3)$$

is determined completely by the initial distribution of X_0 and the transition probabilities $p(x_{i-1}, x_i)$. Since the natural ordering of time has no natural spatial analogue, it is more convenient to work with the symmetric counterpart of (2.2),

$$P(X_i = x_i \mid X_j, j \neq i) = P(X_i = x_i \mid X_{i-1} = x_{i-1}, X_{i+1} = x_{i+1}) \quad (2.4)$$

with obvious modifications for $i = 0, N$. Thus, the conditional behaviour at time i depends only on the states of the chain at the 'neighbour' times $i-1$ and $i+1$. Below, (2.4) is used to motivate the definition of a Markov point process, while (2.3) has an important analogue in the Hammersley-Clifford theorem.

In order to generalise the notion of neighbouring time slots to spatial patterns, let $\sim$ on $\mathcal{X}$ be a symmetric, reflexive relation on $\mathcal{X}$. That is, for any $u, v \in \mathcal{X}$, $u \sim u$ and $u \sim v \Leftrightarrow v \sim u$. The points $u, v \in \mathcal{X}$ are said to be *neighbours* if $u \sim v$.

Example 2.1 Let $(\mathcal{X}, d)$ be a complete, separable metric space. The *fixed range relation* on $\mathcal{X}$ is defined as

$$u \sim v \Leftrightarrow d(u, v) \leq R$$

for some threshold distance R, or equivalently

$$u \sim v \Leftrightarrow B(u, \frac{R}{2}) \cap B(v, \frac{R}{2}) \neq \emptyset$$

where $B(u, R/2)$ denotes the closed ball centred at u with radius $R/2$.

Definition 2.1 The *neighbourhood* $\partial(A)$ of a set $A \subseteq \mathcal{X}$ is defined as

$$\partial(A) = \{x \in \mathcal{X} : x \sim a \text{ for some } a \in A\}.$$

In particular, the neighbourhood of a singleton $A = \{a\}$ contains all neighbours of a,

$$\partial(\{a\}) = \{x \in \mathcal{X} : x \sim a\}.$$

We are now ready to state the definition of a Markov point process in the sense of Ripley and Kelly [181]. It should be noted that the similar concept of a Gibbs point process was already known in statistical physics, see for instance [191, chap. 3] or [165].

Definition 2.2   Let $(\mathcal{X}, d)$ be a complete, separable metric space, $\nu(\cdot)$ a finite, non-atomic Borel measure, and $\pi_\nu(\cdot)$ the distribution of a Poisson process on $\mathcal{X}$ with intensity measure $\nu(\cdot)$.

Let X be a point process on $\mathcal{X}$ specified by means of a density $p(\cdot)$ with respect to $\pi_\nu(\cdot)$. Then X is a *Markov point process* with respect to the symmetric, reflexive relation $\sim$ on $\mathcal{X}$ if for all $\mathbf{x} \in N^f$ such that $p(\mathbf{x}) > 0$,

(a) $p(\mathbf{y}) > 0$ for all $\mathbf{y} \subseteq \mathbf{x}$;

(b) for all $u \in \mathcal{X}$, $p(\mathbf{x} \cup \{u\})/p(\mathbf{x})$ depends only on u and $\partial(\{u\}) \cap \mathbf{x} = \{x \in \mathbf{x} : u \sim x\}$.

Condition (a) states that whenever a configuration $\mathbf{x}$ can occur, so may its subconfigurations. For instance for the hard core model, if no cells with nuclei in $\mathbf{x}$ overlap each other, the same is true for any subset $\mathbf{y} \subseteq \mathbf{x}$. The conditions on the measure $\nu(\cdot)$ imply that X is finite and simple.

For $u \notin \mathbf{x}$, by theorem 1.6, $p(\mathbf{x} \cup \{u\})/p(\mathbf{x}) = \lambda(u; \mathbf{x})$, the Papangelou conditional intensity at u given a configuration $\mathbf{x}$ elsewhere. Hence the local Markov condition (b) is a spatial analogue of (2.4). For $u \in \mathbf{x}$, since the dominating Poisson process is simple, we may set $\lambda(u; \mathbf{x}) = 0$.

Example 2.2   The density of a Poisson process with intensity λ is

$$p(\{x_1, \ldots, x_n\}) = \lambda^n \exp\left[(1 - \lambda)(\mu(\mathcal{X}))\right]$$

with respect to a unit rate Poisson process (cf. example 1.13). Note that $p(\mathbf{x}) > 0$ for all configurations $\mathbf{x}$. Furthermore, the conditional intensity $\lambda(u; \mathbf{x}) = \lambda \, \mathbb{I}\{u \notin \mathbf{x}\}$. Consequently, $p(\cdot)$ is a Markov density for any choice of the neighbourhood relation, in accordance with the fact that a Poisson process represents complete spatial randomness.

Example 2.3 The hard-core model (2.1) is Markov with respect to the neighbourhood relation

$$u \sim v \Leftrightarrow \|u - v\| \leq 2R.$$

To verify conditions (a)–(b) in definition 2.2, note that $p(\mathbf{x}) > 0$ if and only if all members of $\mathbf{x}$ are further than $2R$ apart. Hence whenever $p(\mathbf{x}) > 0$, any subset $\mathbf{y} \subseteq \mathbf{x}$ cannot contain points closer than $2R$ together either, and $p(\mathbf{y}) > 0$. The Papangelou conditional intensity at u given $\mathbf{x} = \{x_1, \ldots, x_n\}$ equals $\lambda(u; \mathbf{x}) = \beta \ \mathbb{I}\{\|u - x_i\| > 2R, i = 1, \ldots n\} = \beta \ \mathbb{I}\{\partial(\{u\}) \cap \mathbf{x} = \emptyset\}$, provided $p(\mathbf{x}) > 0$. We conclude that (2.1) is Markov at range $2R$.

In definition 2.2, condition (b) is a *local Markov property* in the sense that it concerns the behaviour at a single point u given the remainder of the pattern. More generally, the following *spatial Markov property* holds [181].

Theorem 2.1 *Let X be a Markov point process with density $p(\cdot)$ on a complete, separable metric space $(\mathcal{X}, d)$, and consider a Borel set $A \subseteq \mathcal{X}$. Then the conditional distribution of $X \cap A$ given $X \cap A^c$ depends only on X restricted to the neighbourhood*

$$\partial(A) \cap A^c = \{u \in \mathcal{X} \setminus A : u \sim a \text{ for some } a \in A\}.$$

Proof. The proof is based on partitioning a point configuration $\mathbf{z}$ into its restrictions to A and A^c as follows:

$$\mathbf{z} = \mathbf{z}_A \cup \mathbf{z}_{A^c}$$

where $\mathbf{z}_A = \mathbf{z} \cap A$ and $\mathbf{z}_{A^c} = \mathbf{z} \cap A^c$. Fix a configuration $\mathbf{x}$ with $p(\mathbf{x}) > 0$ and consider changing $\mathbf{x}_A$ to $\tilde{\mathbf{x}}_A$. We claim

$$\frac{p(\tilde{\mathbf{x}}_A \cup \mathbf{x}_{A^c})}{p(\mathbf{x}_A \cup \mathbf{x}_{A^c})} = \frac{p(\tilde{\mathbf{x}}_A \cup \mathbf{x}_{\partial(A) \cap A^c})}{p(\mathbf{x}_A \cup \mathbf{x}_{\partial(A) \cap A^c})}. \tag{2.5}$$

To see this, first consider the case $p(\tilde{\mathbf{x}}_A \cup \mathbf{x}_{A^c}) > 0$. Then by Markov property (a), all subsets of $\tilde{\mathbf{x}}_A \cup \mathbf{x}_{A^c}$ also have positive $p(\cdot)$ value. Hence we can factorise the numerator in the left hand side of (2.5) as

$$p(\mathbf{x}_{A^c}) \frac{p(\mathbf{x}_{A^c} \cup \{\tilde{x}_1\})}{p(\mathbf{x}_{A^c})} \frac{p(\mathbf{x}_{A^c} \cup \{\tilde{x}_1, \tilde{x}_2\})}{p(\mathbf{x}_{A^c} \cup \{\tilde{x}_1\})} \cdots \frac{p(\mathbf{x}_{A^c} \cup \tilde{\mathbf{x}}_A)}{p(\mathbf{x}_{A^c} \cup \{\tilde{x}_1, \ldots, \tilde{x}_{m-1}\})}$$

where $\tilde{\mathbf{x}}_A = \{\tilde{x}_1, \ldots, \tilde{x}_m\}$. By the local Markov property, in each of the quotient terms above, $\mathbf{x}_{A^c}$ can be replaced by $\mathbf{x}_{\partial(A) \cap A^c}$. The same argument applied to the denominator in the left hand side of (2.5) proves (2.5).

Secondly, if $p(\tilde{\mathbf{x}}_A \cup \mathbf{x}_{A^c}) = 0$, consider the first zero-term in the above factorisation, say

$$\frac{p(\mathbf{x}_{A^c} \cup \{\tilde{x}_1, \ldots, \tilde{x}_i\})}{p(\mathbf{x}_{A^c} \cup \{\tilde{x}_1, \ldots, \tilde{x}_{i-1}\})} = 0.$$

As before, replace $\mathbf{x}_{A^c}$ by $\mathbf{x}_{\partial(A) \cap A^c}$ to conclude $p(\mathbf{x}_{\partial(A) \cap A^c} \cup \{\tilde{x}_1, \ldots, \tilde{x}_i\}) = 0$, hence $p(\tilde{\mathbf{x}}_A \cup \mathbf{x}_{\partial(A) \cap A^c}) = 0$. $\qquad\square$

Let us briefly return to the microscopy example discussed in the beginning of this section. Here, the spatial Markov property amounts to saying that the pattern observed in the window of the microscope is conditionally independent of those nuclei in the tissue further than $2R$ from the window boundary given those closer to the window (that might intersect a cell having its nucleus inside).

2.2   The Hammersley–Clifford theorem

The goal of this section is to find a factorisation formula similar in spirit to (2.3). The spatial analogues of the transition probabilities of a Markov chain are the so-called *clique interaction functions*.

Definition 2.3   Let $\sim$ be a symmetric, reflexive neighbourhood relation on $\mathcal{X}$. A configuration $\mathbf{x} \in N^f$ is said to be a *clique* if all members of $\mathbf{x}$ are each other's neighbour, i.e. for each $u, v \in \mathbf{x}$, $u \sim v$. By convention, the empty set is a clique as well.

The Hammersley–Clifford theorem [181] provides an explicit factorisation of the density of a Markov point process in terms of interactions between the points.

Theorem 2.2   *A point process density $p : N^f \to [0, \infty)$ is Markov with respect to a neighbourhood relation $\sim$ if and only if it there is a measurable function $\phi : N^f \to [0, \infty)$ such that*

$$p(\mathbf{x}) = \prod_{\text{cliques } \mathbf{y} \subseteq \mathbf{x}} \phi(\mathbf{y})$$

for all $\mathbf{x} \in N^f$.

The product over cliques in the Hammersley–Clifford factorisation above may be replaced by a product over all subsets of $\mathbf{x}$ by setting $\phi(\mathbf{y}) = 1$ whenever $\mathbf{y}$ is not a clique.

Proof.   Suppose that $p(\mathbf{x}) = \prod_{\text{cliques } \mathbf{y} \subseteq \mathbf{x}} \phi(\mathbf{y})$, for all $\mathbf{x} \in N^f$. In order to show that $p(\cdot)$ is Markov, we need to check conditions (a) and (b) of definition 2.2. To verify (a), suppose $p(\mathbf{x}) \neq 0$. Then $\phi(\mathbf{y}) \neq 0$ for all cliques $\mathbf{y} \subseteq \mathbf{x}$. If $\mathbf{z} \subseteq \mathbf{x}$, a fortiori $\phi(\mathbf{y}) \neq 0$ for any clique $\mathbf{y} \subseteq \mathbf{z}$, and therefore

$$p(\mathbf{z}) = \prod_{\text{cliques } \mathbf{y} \subseteq \mathbf{z}} \phi(\mathbf{y}) > 0.$$

As for (b), let $\mathbf{x} \in N^f$ be a configuration such that $p(\mathbf{x}) > 0$, and $u \in \mathcal{X}$. Then,

$$\frac{p(\mathbf{x} \cup \{u\})}{p(\mathbf{x})} = \frac{\prod_{\text{cliques } \mathbf{y} \subseteq \mathbf{x}} \phi(\mathbf{y}) \prod_{\text{cliques } \mathbf{y} \subseteq \mathbf{x}} \phi(\mathbf{y} \cup \{u\})}{\prod_{\text{cliques } \mathbf{y} \subseteq \mathbf{x}} \phi(\mathbf{y})}$$

$$= \prod_{\text{cliques } \mathbf{y} \subseteq \mathbf{x}} \phi(\mathbf{y} \cup \{u\}).$$

Since the interaction function $\phi(\mathbf{y} \cup \{u\}) = 1$ whenever $\mathbf{y} \cup \{u\}$ is not a clique, the conditional intensity $\lambda(u; \mathbf{x})$ depends only on u and its neighbours in $\mathbf{x}$.

Conversely, suppose $p(\cdot)$ is a Markov density. Define an interaction function $\phi : N^f \to [0, \infty)$ inductively by

$$\phi(\emptyset) = \alpha = p(\emptyset)$$
$$\phi(\mathbf{x}) = 1 \qquad \text{if } \mathbf{x} \text{ is not a clique}$$
$$\phi(\mathbf{x}) = \frac{p(\mathbf{x})}{\prod_{\mathbf{y}:\mathbf{x}\neq\mathbf{y}\subset\mathbf{x}} \phi(\mathbf{y})} \qquad \text{if } \mathbf{x} \text{ is a clique}$$

with the convention $0/0 = 0$. Note that if $\prod_{\mathbf{y}:\mathbf{x}\neq\mathbf{y}\subset\mathbf{x}} \phi(\mathbf{y}) = 0$ necessarily $p(\mathbf{y}) = 0$ for some $\mathbf{y}$, and therefore $p(\mathbf{x}) = 0$. Hence $\phi(\cdot)$ is well-defined. To show that $p(\cdot)$ has the required product form, we use induction on the number of points. By definition the factorisation holds for the empty set. Assume the factorisation holds for configurations with up to $n - 1$ points, and consider a pattern $\mathbf{x}$ of cardinality $n \geq 1$. We will distinguish three cases.

- Firstly, suppose that $\mathbf{x}$ is not a clique, and $p(\mathbf{x}) = 0$. Then there exist $v, w \in \mathbf{x}$ such that $v \not\sim w$. Furthermore, assume $\prod_{\mathbf{y}:\mathbf{x}\neq\mathbf{y}\subset\mathbf{x}} \phi(\mathbf{y}) > 0$. By the induction hypothesis, $p(\mathbf{y}) > 0$ for all proper subsets $\mathbf{y}$ of $\mathbf{x}$, hence

$$0 = \frac{p(\mathbf{x} = \mathbf{z} \cup \{w, v\})}{p(\mathbf{z} \cup \{v\})} p(\mathbf{z} \cup \{v\}) = \frac{p(\mathbf{z} \cup \{w\})}{p(\mathbf{z})} p(\mathbf{z} \cup \{v\}) > 0$$

  where $\mathbf{z} = \mathbf{x} \setminus \{w, v\}$. The assumption $\prod_{\mathbf{y}:\mathbf{x}\neq\mathbf{y}\subset\mathbf{x}} \phi(\mathbf{y}) > 0$ leads to a contradiction, hence $\prod_{\mathbf{y}:\mathbf{x}\neq\mathbf{y}\subset\mathbf{x}} \phi(\mathbf{y}) = 0 = p(\mathbf{x})$.
- Next, let $\mathbf{x}$ be a clique for which $p(\mathbf{x}) = 0$. Then, $\phi(\mathbf{x}) = 0$ by definition and hence $p(\mathbf{x}) = \prod_{\mathbf{y}\subset\mathbf{x}} \phi(\mathbf{y})$.
- Finally, consider the case that $p(\mathbf{x}) > 0$. If $\mathbf{x}$ is a clique, $p(\mathbf{x}) = \phi(\mathbf{x}) \prod_{\mathbf{y}:\mathbf{x}\neq\mathbf{y}\subset\mathbf{x}} \phi(\mathbf{y}) = \prod_{\mathbf{y}\subset\mathbf{x}} \phi(\mathbf{y})$. If $\mathbf{x}$ is not a clique, write $\mathbf{x} = \mathbf{z}\cup\{v, w\}$, for some $v \not\sim w$. Since $p(\cdot)$ is a Markov density, $p(\mathbf{z}) > 0$, thus

$$p(\mathbf{x}) = \frac{p(\mathbf{z} \cup \{v, w\})}{p(\mathbf{z} \cup \{v\})} p(\mathbf{z} \cup \{v\}) = \frac{p(\mathbf{z} \cup \{w\})}{p(\mathbf{z})} p(\mathbf{z} \cup \{v\})$$

$$= \prod_{\mathbf{y}\subseteq\mathbf{z}} \phi(\mathbf{y} \cup \{w\}) \prod_{\mathbf{y}\subseteq\mathbf{z}\cup v} \phi(\mathbf{y}) = \prod_{\text{cliques } \mathbf{y}\subseteq\mathbf{x}} \phi(\mathbf{y})$$

using the fact that $\phi(\mathbf{y}) = 1$ for any $\mathbf{y}$ containing both v and w. $\qquad\square$

The Hammersley–Clifford theorem is useful for breaking up a high-dimensional joint distribution into manageable clique interaction functions that are easier to interpret and have a lower dimension.

Example 2.4 Let $\mathcal{X}$ be a compact subset of $\mathbb{R}^d$ with positive volume $\mu(\mathcal{X})$, and consider a Poisson process of rate λ on $\mathcal{X}$. Then

$$p(\mathbf{x}) = e^{(1-\lambda)\mu(\mathcal{X})} \prod_{x \in \mathbf{x}} \lambda.$$

Hence the interaction function is given by $\phi(\emptyset) = \exp[(1 - \lambda)\mu(\mathcal{X})]$ for the empty set, and $\phi(\{u\}) = \lambda$ for $u \in \mathcal{X}$. For sets with two or more points, $\phi \equiv 1$, confirming the lack of interaction between points.

Example 2.5 For the hard core model (2.1), $\phi(\emptyset) = \alpha$, $\phi(\{u\}) = \beta$ $(u \in \mathcal{X})$ and

$$\phi(\{u, v\}) = \begin{cases} 1 & \text{if } \|u - v\| > 2R \\ 0 & \text{otherwise} \end{cases}.$$

The interaction function is identically 1 on configurations consisting of three or more points.

Example 2.6 A special class of Markov point processes is that containing the *pairwise interaction* models specified by a density of the form [46]

$$p(\mathbf{x}) = \alpha \prod_{x \in \mathbf{x}} \beta(x) \prod_{u,v \in \mathbf{x}: u \sim v} \gamma(u, v) \qquad (2.6)$$

where $\alpha > 0$ is the normalising constant, $\beta : \mathcal{X} \to [0, \infty)$ the intensity function and $\gamma : \mathcal{X} \times \mathcal{X} \to [0, \infty)$ the pair interaction function. In statistical physics terminology the model (2.6) is also referred to as a *pair potential process* [176].

It is easily seen that the hard-core model (2.1) is a pairwise interaction process at range $2R$ with $\beta(x) \equiv \beta$, and $\gamma(u, v) = 1\{u \not\sim v\}$. Taking $\gamma \equiv 1$, yields a Poisson process.

The Hammersley–Clifford theorem can be used when designing new models; instead of defining a joint density $p(\mathbf{x})$ for all point patterns $\mathbf{x}$, it is sufficient to define the interaction functions $\phi(\cdot)$. When it is plausible that interactions occur only between pairs of neighbours, a model can be specified quite simply by giving the $\beta(\cdot)$ and $\gamma(\cdot,\cdot)$ functions in (2.6). Some care is needed though; one has to make sure that a particular choice of interaction functions results in a $p(\cdot)$ that is integrable!

2.3   Markov marked point processes

Marked point processes are point processes in which a mark is attached to each event. As such, they are often useful in applications where measurements are made at each observation, or where the points are of different types. For instance, forest researchers mapping a pattern of trees may record the height or stem diameter for each of the trees, or if several species are present, the mark may be a species label; in material science or image analysis, a pattern of objects may be modelled as a point process of nuclei (e.g. centre of gravity or other typical point) marked by shape descriptors.

The definition of a Markov point process readily extends to marked point processes [11; 181]. Let $\mathcal{X}$ and $\mathcal{K}$ be complete, separable metric spaces, and $m(\cdot)$ a probability distribution on the Borel σ-algebra of $\mathcal{K}$. An appropriate dominating Poisson process in this context has locations following a Poisson process on $\mathcal{X}$ marked independently by m-distributed labels.

Definition 2.4   Let $(\mathcal{X}, d)$ and $(\mathcal{K}, d')$ be complete, separable metric spaces, $\nu(\cdot)$ a finite, non-atomic Borel measure on $\mathcal{X}$, $m(\cdot)$ a probability distribution on the Borel σ-algebra of $\mathcal{K}$, and $\pi_{\nu \times m}(\cdot)$ the distribution of a Poisson process on $\mathcal{X} \times \mathcal{K}$ with intensity measure $\nu \times m$.

Let Y be a marked point process with positions in $\mathcal{X}$ and marks in $\mathcal{K}$ specified by means of a density $p(\cdot)$ with respect to $\pi_{\nu \times m}$. Then Y is a *Markov marked point process* with respect to the symmetric, reflexive relation $\sim$ on $\mathcal{X} \times \mathcal{K}$ if for all $\mathbf{y}$ such that $p(\mathbf{y}) > 0$,

(a) $p(\mathbf{z}) > 0$ for all $\mathbf{z} \subseteq \mathbf{y}$;
(b) for all $(u, l) \in \mathcal{X} \times \mathcal{K}$, $p(\mathbf{y} \cup \{(u, l)\})/p(\mathbf{y})$ depends only on (u, l) and
$\partial(\{(u, l)\}) \cap \mathbf{y} = \{(x, k) \in \mathbf{y} : (u, l) \sim (x, k)\}$.

The Hammersley–Clifford theorem is valid for marked point processes as well [11; 181]. Thus, a probability density $p(\cdot)$ defines a Markov marked point process with respect to $\sim$ on $\mathcal{X} \times \mathcal{K}$ if and only if it can be factorised as

$$p(\mathbf{y}) = \prod_{\text{cliques } \mathbf{z} \subseteq \mathbf{y}} \phi(\mathbf{z}) \tag{2.7}$$

for all $\mathbf{y} \in N^{\text{f}}$, where the product is restricted to $\sim$–cliques $\mathbf{z} \subseteq \mathbf{y}$, and $\phi : N^{\text{f}} \to [0, \infty)$ is the interaction function.

Example 2.7 Let X be a finite Markov point process on $\mathcal{X}$ with respect to a neighbourhood relation $\sim$. Write $p_X(\cdot)$ for the density of X with respect to a Poisson process on $\mathcal{X}$ with intensity measure $\nu(\cdot)$. Suppose to each of the points of X a mark is attached independently according to a probability distribution $m(\cdot)$ on the mark space $\mathcal{K}$. Then the probability distribution governing the total number of points is given by (cf. section 1.4)

$$p_n = \frac{e^{-\nu(\mathcal{X})}}{n!} \int_{\mathcal{X}} \cdots \int_{\mathcal{X}} p_X(\{x_1, \ldots x_n\}) \, d\nu(x_1) \cdots d\nu(x_n), \quad n \in \mathbb{N}_0;$$

conditionally on $\{N(\mathcal{X} \times \mathcal{K}) = n\}$, the n points are distributed according to

$$\frac{p_X(\{x_1, \ldots, x_n\}) \, d\nu(x_1) \cdots d\nu(x_n) \, dm(k_1) \cdots dm(k_n)}{\int_{\mathcal{X}} \cdots \int_{\mathcal{X}} p_X(\{x_1, \ldots x_n\}) \, d\nu(x_1) \cdots d\nu(x_n)}.$$

Hence, if $f(\{(x_1, k_1), \ldots, (x_n, k_n)\}) = \{x_1, \ldots x_n\}$ denotes the projection on the first component, the marked point process has density $p(\mathbf{y}) = p_X(f(\mathbf{y}))$ with respect to a Poisson process with intensity measure $\nu \times m$ on $\mathcal{X} \times \mathcal{K}$. Consequently, the conditional intensity of the marked process Y at $\{(u, k)\}$ equals the conditional intensity of X at u, and the marked point process is Markov with respect to the mark-independent neighbourhood relation $\sim_2$ on $\mathcal{X} \times \mathcal{K}$ defined by $(u, k) \sim_2 (v, l) \Leftrightarrow u \sim v$.

Example 2.8 Let X_1, X_2 be two independent, homogeneous Poisson processes on a compact window $\mathcal{X} \subseteq \mathbb{R}^2$ with intensity $\beta > 0$, and condition on the event E that no point of one component is closer than R to a point of the other component. The resulting bivariate point process Y is called the *penetrable spheres mixture model* [81; 186; 187; 221]. A typical realisation for $\beta = 50$ and $R = 0.1$ is depicted in Figure 2.2.

Clearly, Y is absolutely continuous with respect to the product of two unit intensity Poisson processes, with density

$$p(\mathbf{x}_1, \mathbf{x}_2) = \alpha \, \beta^{n(\mathbf{x}_1)+n(\mathbf{x}_2)} \, \mathbb{I}\{d(\mathbf{x}_1, \mathbf{x}_2) > R\}$$

where $\alpha > 0$ is the normalising constant, $n(\mathbf{z})$ denotes the number of points in $\mathbf{z}$, and $d(\mathbf{x}_1, \mathbf{x}_2)$ is the smallest distance between a point of $\mathbf{x}_1$ and one of $\mathbf{x}_2$. If either component is empty, $d(\mathbf{x}_1, \mathbf{x}_2)$ is set to infinity. The Papangelou conditional intensity for adding a new $u \in \mathcal{X}$ to the first component of a bivariate pattern $(\mathbf{x}_1, \mathbf{x}_2)$ that respects the hard core distance between its components is

$$\beta \, \mathbb{I}\{d(u, \mathbf{x}_2) > R\}$$

which depends only on points in $\mathbf{x}_2$ within distance R of u. A similar argument applies when a point is added to the second component. Since moreover $\{(\mathbf{x}_1, \mathbf{x}_2) : d(\mathbf{x}_1, \mathbf{x}_2) > R\}$ is hereditary, seen as a marked point process as in section 1.2, Y is Markov with respect to the relation

$$(u, k) \sim (v, l) \Leftrightarrow k \neq l \text{ and } \|u - v\| \leq R$$

for $u, v \in \mathcal{X}$ and $k, l \in \{1, 2\}$. A fortiori, Y is Markov at range R.

Example 2.9 Let $\mathcal{X}$ be a compact window in $\mathbb{R}^2$, and consider the *marked pairwise interaction process* on $\mathcal{X} \times \{1, \ldots, M\}$ defined by [11]

$$p(\mathbf{y}) = \alpha \prod_{(x,k)\in\mathbf{y}} \beta_k \prod_{(u,k),(v,l)\in\mathbf{y}} \gamma_{kl}(\|u - v\|)$$

for some $\alpha > 0$, intensity parameters $\beta_k > 0$, and measurable interaction functions $\gamma_{kl} : [0, \infty) \to [0, \infty)$. The reference distribution is that of a unit rate Poisson process of locations marked by a uniformly distributed label.

Without loss of generality, assume that $\gamma_{kl} = \gamma_{lk}$. Then, if $\gamma_{kl}(r) = 1$ for $r > r_{kl}$ for some range parameters $r_{kl} > 0$, the process is Markov with respect to the mark dependent relation

$$(u, k) \sim (v, l) \Leftrightarrow \|u - v\| \leq r_{kl}.$$

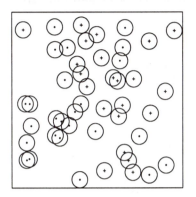

Fig. 2.2 Realisation of a penetrable spheres mixture model on the unit square with intensity parameter $\beta = 50$. The two components are indicated by •, +. Around each of the points, a disc of radius 0.05 is drawn.

To see this, let $p(\mathbf{y}) > 0$ for some marked point pattern $\mathbf{y} = \{(x_1, k_1),$ $\dots, (x_n, k_n)\}$. Then all $\gamma_{k_i k_j}(\|x_i - x_j\|)$ are strictly positive, hence $p(\mathbf{z})$ is strictly positive for any $\mathbf{z} \subseteq \mathbf{y}$. Regarding the local Markov property, note that

$$\lambda((u, l); \mathbf{y}) = \frac{p(\mathbf{y} \cup \{(u, l)\})}{p(\mathbf{y})} = \beta_l \prod_{i=1}^{n} \gamma_{lk_i}(\|u - x_i\|).$$

Since $\gamma_{lk_i}(\|u - x_i\|) = 1$ whenever $\|u - x_i\| > r_{lk_i}$, $\lambda((u, l); \mathbf{y})$ depends only on (u, l) and its neighbours in $\mathbf{y}$. Hence $p(\cdot)$ is Markov with respect to the above relation. A fortiori, the density $p(\cdot)$ is Markov at range $r = \max_{k,l} r_{kl}$.

The penetrable spheres mixture model can be cast in a pairwise interaction framework by setting

$$\gamma_{12}(u, v) = \mathrm{I\!I}\{\|u - v\| > R\}$$

and $\gamma_{11} \equiv 1$. More generally, a *soft core model* has $\gamma_{kl} \equiv \gamma_{kl}$ for some $\gamma_{kl} \in (0, 1)$. The latter model does not actually prohibit R-close pairs, but discourages them with a force determined by γ_{kl}.

In example 2.7, we noted that for independently marked points, the Markov structure of the underlying point process is carried over to the marked point process. This is not the case in general.

Example 2.10   Consider the penetrable spheres mixture model of example 2.8. The point process $X = X_1 \cup X_2$ has density (with respect to a unit rate Poisson process on $\mathcal{X}$)

$$
\begin{aligned}
p(\mathbf{x}) &= e^{-\mu(\mathcal{X})} \sum p(\mathbf{x}_1, \mathbf{x}_2) \\
&= e^{-\mu(\mathcal{X})} \sum \alpha \, \beta^{n(\mathbf{x}_1)} \, \beta^{n(\mathbf{x}_2)} \, \mathbb{I}\{d(\mathbf{x}_1, \mathbf{x}_2) > R\} \\
&= \alpha e^{-\mu(\mathcal{X})} \, \beta^{n(\mathbf{x})} \, k(\mathbf{x})
\end{aligned}
$$

where the sum is over all ordered partitions of $\mathbf{x}$ into two groups $\mathbf{x}_1$ and $\mathbf{x}_2$, $n(\mathbf{z})$ is the cardinality of $\mathbf{z}$, and $k(\mathbf{x})$ denotes the number of partitions such that the hard core distance between the two groups is respected. In more geometric terms (see Figure 2.2), if a ball $B(x, R/2)$ of radius $R/2$ is placed around each of the points $x \in \mathbf{x}$, $k(\mathbf{x})$ is the number of partitions such that

$$
(\cup_{\eta \in \mathbf{x}_1} B(\eta, R/2)) \cap (\cup_{\zeta \in \mathbf{x}_2} B(\zeta, R/2)) = \emptyset.
$$

It follows that $k(\mathbf{x}) = 2^{c(\mathbf{x})}$, where $c(\mathbf{x})$ is the number of connected components in $\cup_{x \in \mathbf{x}} B(x, R/2)$. Hence

$$
p(\mathbf{x}) = \alpha \, e^{-\mu(\mathcal{X})} \, \beta^{n(\mathbf{x})} \, 2^{c(\mathbf{x})},
$$

the density of a *continuum random cluster model* [107; 143]. Finally, note that the conditional intensity

$$
\lambda(u; \mathbf{x}) = \beta \, 2^{c(\mathbf{x} \cup \{u\}) - c(\mathbf{x})}
$$

may depend on points far away from u.

Next, restrict attention to finite mark spaces, so that we can define the component point processes X_i consisting of those points that have mark

$i \in \mathcal{K}$. It is natural to ask, whether any Markov property of the marked point process is inherited by the components.

Example 2.11 Further to example 2.8, note that the first component X_1 has density

$$
\begin{aligned}
p(\mathbf{x_1}) &= \int_{N^f} \alpha \, \beta^{n(\mathbf{x_1})} \, \beta^{n(\mathbf{x_2})} \, \mathbb{I}\{d(\mathbf{x_1}, \mathbf{x_2}) > R\} \, d\pi(\mathbf{x_2}) \\
&= \int_{N^f} \alpha \, \beta^{n(\mathbf{x_1})} \, \beta^{n(\mathbf{x_2})} \, \mathbb{I}\{\mathbf{x_2} \cap U_{\mathbf{x_1}} = \emptyset\} \, d\pi(\mathbf{x_2}) \\
&= \alpha \, e^{-(1-\beta)\mu(\mathcal{X})} \, \beta^{n(\mathbf{x_1})} \, e^{-\beta\mu(U_{\mathbf{x_1}} \cap \mathcal{X})}
\end{aligned}
$$

where $U_{\mathbf{x}} = \bigcup_{x \in \mathbf{x_1}} B(x, R)$ denotes the union of balls around points in $\mathbf{x_1}$, and $\pi(\cdot)$ is the reference Poisson process distribution. Thus, the process may be called an *area-interaction process* [8]. The conditional intensity

$$
\lambda(u; \mathbf{x_1}) = \beta \, \exp\left[-\beta\mu(\mathcal{X} \cap (B(u, R) \setminus U_{\mathbf{x_1}}))\right], \qquad u \notin \mathbf{x_1}
$$

depends only on those points $x \in \mathbf{x_1}$ that are closer than $2R$ to u. By symmetry, the same property holds for the second component. Summarising, the component processes are Markov at range $2R$, while the marked point process itself is Markov at range R.

Example 2.12 Let $Y = (X_1, X_2)$ be a bivariate pairwise interaction point process on a compact window $\mathcal{X}$ in $\mathbb{R}^2$ defined by its density

$$
p(\mathbf{x_1}, \mathbf{x_2}) = \alpha \, \beta_1^{n(\mathbf{x_1})} \, \beta_2^{n(\mathbf{x_2})} \prod_{u, v \in \mathbf{x_1} \cup \mathbf{x_2}} \gamma_{m(u) \, m(v)}(\|u - v\|)
$$

with respect to the product of two unit rate Poisson processes. Here $m(u)$ denotes the component of u. As in example 2.9, we assume that $\gamma_{kl}(r) = 1$ for $r > r_{kl}$. Writing $\pi(\cdot)$ for the distribution of a unit rate Poisson process and integrating over the second component yields

$$
p(\mathbf{x_1}) = \alpha \, \beta_1^{n(\mathbf{x_1})} \prod_{u, v \in \mathbf{x_1}} \gamma_{11}(\|u - v\|) \, I_2(\mathbf{x_1})
$$

where $n(\cdot)$ denotes cardinality and

$$
I_2(\mathbf{x_1}) = \int_{N^f} \beta_2^{n(\mathbf{x_2})} \prod_{u, v \in \mathbf{x_2}} \gamma_{22}(\|u - v\|) \prod_{u \in \mathbf{x_1}, v \in \mathbf{x_2}} \gamma_{12}(\|u - v\|) \, d\pi(\mathbf{x_2}).
$$

The integral $I_2(\mathbf{x}_1)$ in general does not factorise, hence X_1 is not necessarily a Markov point process at range r_{1i} (nor at range $\max_{k,l\in\{1,2\}} r_{kl}$).

The above examples show that, in general, the components of a multivariate point process do not inherit Markovianity.

Fig. 2.3   Neighbourhood graph on $\mathbf{x} = \{a, b, c, d\}$.

Reversely, suppose a multivariate point process Y has independent components that are Markov with respect to some relation $\sim$ on $\mathcal{X}$. Then clearly, interpreted as a marked point process, Y is Markov with respect to the mark dependent relation $(u, k) \sim (v, l)$ whenever $u \sim v$ and $k = l$. The unmarked point process X defined by disregarding the types however is not necessarily Markov with respect to $\sim$.

Example 2.13   Let $\sim$ be a neighbourhood relation on $\mathcal{X}$. Define $Y = (X_1, X_2)$, where X_1 and X_2 are independent, identically distributed *Strauss processes* [101; 210] on $\mathcal{X}$ having density

$$p(\mathbf{x}) = \alpha\,\gamma^{s(\mathbf{x})}.$$

Here $s(\mathbf{x})$ denotes the number of distinct neighbour pairs in $\mathbf{x}$ and $\gamma \in (0, 1)$ is the interaction parameter. Thus, configurations with many neighbouring points are discouraged. If $\sim$ is a fixed range relation as in example 2.1, $p(\cdot)$ is also known as the *soft core model.* Since $s(\{x_1, \ldots, x_n\}) = \sum_{1=i<j\leq n} \mathbb{I}\{d(x_i, x_j) \leq R\}$ is a permutation invariant, Borel measurable function on $\mathcal{X}^n$, the soft core density is measurable on N^f.

Define $X = X_1 \cup X_2$, the unmarked point process associated with Y. Suppose the space $\mathcal{X}$ is rich enough to allow for a configuration $\mathbf{x} = \{a, b, c, d\}$ such that $a \sim b \sim c \sim d$ are the only related points (see Figure 2.3). Then the Papangelou conditional intensity of X at d given $\mathbf{x} \setminus \{d\}$ is

$$
\begin{aligned}
\lambda(d; \{a, b, c\}) &= \frac{\sum_{\mathbf{y} \subseteq \mathbf{x}} p(\mathbf{y}) \, p(\mathbf{x} \setminus \mathbf{y})}{\sum_{\mathbf{y} \subseteq \mathbf{x} \setminus \{d\}} p(\mathbf{y}) \, p(\{a, b, c\} \setminus \mathbf{y})} = \frac{\gamma^4 + 6\gamma^2 + 1}{\gamma^2 + 2\gamma + 1} \\
&\neq \frac{\gamma^2 + 2\gamma + 1}{2} = \lambda(d; \{a, c\}).
\end{aligned}
$$

It follows that X is not a Markov point process with respect to the given relation $\sim$.

Perhaps surprisingly, one has to consider sets of four points in the above counterexample; the pair and triple interaction functions do respect the neighbourhood relation [119].

Lemma 2.1 *Let X_1 and X_2 be independent and identically distributed Markov point processes with respect to a reflexive, symmetric relation $\sim$ on a complete, separable metric space $(\mathcal{X}, d)$, defined on a common underlying measure space. Let $X = X_1 \cup X_2$ be the superposition of X_1 and X_2. Then, if X_1 has density $p(\cdot)$ with respect to a Poisson process on $\mathcal{X}$ with finite, non-atomic intensity measure $\nu(\cdot)$, X is absolutely continuous with density*

$$
p_s(\mathbf{x}) = e^{-\nu(\mathcal{X})} \sum_{\mathbf{x}_1, \mathbf{x}_2} p(\mathbf{x}_1) \, p(\mathbf{x}_2)
$$

where the sum ranges over all partitions of $\mathbf{x}$ into two components. Moreover, $p_s(\mathbf{x})$ factorises into a product of clique interaction functions for all configurations $\mathbf{x}$ with at most three members.

Proof. The formula for the superposition density is a direct result of the independence of the components. Suppose $p_s(\mathbf{x})$ is strictly positive for some configuration $\mathbf{x}$. Then a partition $\mathbf{x}_1 \cup \mathbf{x}_2 = \mathbf{x}$ exists for which $p(\mathbf{x}_1) > 0$ and $p(\mathbf{x}_2) > 0$. Since the density of a Markov point process is hereditary, $p(\mathbf{x}_i \cap \mathbf{y})$, $i \in \{1,2\}$, is strictly positive for every configuration $\mathbf{y} \subseteq \mathbf{x}$. Therefore

$$p_s(\mathbf{y}) \geq e^{-\nu(\mathcal{X})} p_1(\mathbf{x}_1 \cap \mathbf{y}) p_2(\mathbf{x}_2 \cap \mathbf{y}) > 0$$

for all $\mathbf{y} \subseteq \mathbf{x}$, hence $p_s(\cdot)$ is hereditary.

Since X_1 (and X_2) is Markov with respect to $\sim$, by the Hammersley–Clifford theorem 2.2,

$$p_s(\mathbf{x}) = e^{-\nu(\mathcal{X})} \sum_{\mathbf{x}_1, \mathbf{x}_2} \left[\prod_{\mathbf{y}_1 \subseteq \mathbf{x}_1} \phi(\mathbf{y}_1) \prod_{\mathbf{y}_2 \subseteq \mathbf{x}_2} \phi(\mathbf{y}_2) \right]$$

for some interaction function $\phi(\cdot)$ satisfying $\phi(\mathbf{z}) = 1$ unless $\mathbf{z}$ is a $\sim$-clique. Set $\phi_s(\emptyset) = p_s(\emptyset) = e^{-\nu(\mathcal{X})} \phi(\emptyset)^2$. Then

$$p_s(\{\xi\}) = 2 e^{-\nu(\mathcal{X})} \phi(\emptyset)^2 \phi(\{\xi\}) = 2 \phi_s(\emptyset) \phi(\{\xi\})$$

and hence $\phi_s(\{\xi\}) = 2 \phi(\{\xi\})$. Regarding doublets,

$$p_s(\{\xi, \eta\}) = \phi_s(\emptyset) \phi_s(\{\xi\}) \phi_s(\{\eta\}) + 2\phi_s(\emptyset) \phi(\{\xi\}) \phi(\{\eta\})(\phi(\{\xi, \eta\}) - 1)$$

and therefore

$$\phi_s(\{\xi, \eta\}) = 1 + \frac{1}{2}(\phi(\{\xi, \eta\}) - 1),$$

provided $\phi(\{\xi\})$ and $\phi(\{\eta\})$ are non-zero. Now $\{\xi, \eta\}$ is not a clique if and only if $\xi \not\sim \eta$. In this case $\phi(\{\xi, \eta\}) = 1$, hence $\phi_s(\{\xi, \eta\}) = 1$ as well. If either $\phi(\{\xi\}) = 0$ or $\phi(\{\eta\}) = 0$, then, since both $p(\cdot)$ and $p_s(\cdot)$ are hereditary, $p(\{\xi, \eta\}) = p_s(\{\xi, \eta\}) = 0$ and we set $\phi_s(\{\xi, \eta\}) = 0$.

Finally,

$$p_s(\{\xi, \eta, \zeta\}) = \phi_s(\emptyset) \phi(\{\xi\}) \phi(\{\eta\}) \phi(\{\zeta\}) [$$
$$2\phi(\{\xi, \eta\}) \phi(\{\xi, \zeta\}) \phi(\{\eta, \zeta\}) \phi(\{\xi, \eta, \zeta\}) +$$
$$2\phi(\{\xi, \eta\}) + 2\phi(\{\xi, \zeta\}) + 2\phi(\{\eta, \zeta\})]$$

hence

$$
\begin{aligned}
\phi_s(\{\xi,\eta,\zeta\}) &= \frac{1}{4}\frac{\phi(\{\xi,\eta\})\,\phi(\{\xi,\zeta\})\,\phi(\{\eta,\zeta\})}{\phi_s(\{\xi,\eta\})\,\phi_s(\{\xi,\zeta\})\,\phi_s(\{\eta,\zeta\})}\,\phi(\{\xi,\eta,\zeta\}) \\
&\quad + \frac{1}{4}\frac{\phi(\{\xi,\eta\})+\phi(\{\xi,\zeta\})+\phi(\{\eta,\zeta\})}{\phi_s(\{\xi,\eta\})\,\phi_s(\{\xi,\zeta\})\,\phi_s(\{\eta,\zeta\})} \\
&= 1 + \frac{1}{4}\frac{\phi(\{\xi,\eta\})\,\phi(\{\xi,\zeta\})\,\phi(\{\eta,\zeta\})}{\phi_s(\{\xi,\eta\})\,\phi_s(\{\xi,\zeta\})\,\phi_s(\{\eta,\zeta\})}\,(\phi(\{\xi,\eta,\zeta\})-1) \\
&\quad + \frac{1}{8}\frac{(\phi(\{\xi,\eta\})-1)\,(\phi(\{\xi,\zeta\})-1)\,(\phi(\{\eta,\zeta\})-1)}{\phi_s(\{\xi,\eta\})\,\phi_s(\{\xi,\zeta\})\,\phi_s(\{\eta,\zeta\})},
\end{aligned}
$$

assuming the numerator is non-zero (and $\phi_s(\{\xi,\eta,\zeta\}) = 0$ otherwise). If $\{\xi,\eta,\zeta\}$ is not a clique, as X_1 and X_2 are Markov, $\phi(\{\xi,\eta,\zeta\}) = 1$. Furthermore, there must be a pair in $\{\xi,\eta,\zeta\}$ that are not neighbours and hence at least one of $\phi(\{\xi,\eta\})$, $\phi(\{\xi,\zeta\})$, $\phi(\{\eta,\zeta\})$ must be 1. In summary,

$$
p_s(\mathbf{x}) = \prod_{\text{cliques } \mathbf{y} \subseteq \mathbf{x}} \phi_s(\mathbf{y})
$$

provided the cardinality $n(\mathbf{x}) \le 3$. $\qquad\square$

When the component processes are not identically distributed, lemma 2.1 is no longer valid [119].

Example 2.14 Let $X = X_1 \cup X_2$ be the superposition of two independent Strauss processes as in example 2.13 with parameter values $\gamma_1 \ne \gamma_2$. Suppose the space $\mathcal{X}$ is rich enough to allow for a configuration $\{a,b,c\}$ such that $a \sim b \sim c$ but $a \not\sim c$. Then, in the notation of the lemma above, $\phi_s(\{a\}) = \phi_s(\{b\}) = \phi_s(\{c\}) = 2$, $\phi_s(\{a,b\}) = \phi_s(\{b,c\}) = \frac{1}{4}(\gamma_1 + \gamma_2 + 2)$, and $\phi_s(\{a,c\}) = 1$. Now,

$$
p_s(\{a,b,c\}) = \phi_s(\emptyset)\left((\gamma_1+1)^2 + (\gamma_2+1)^2\right),
$$

hence

$$
\phi_s(\{a,b,c\}) = 2\frac{(\gamma_1+1)^2 + (\gamma_2+1)^2}{(\gamma_1+\gamma_2+2)^2}. \tag{2.8}
$$

The set $\{a,b,c\}$ is no clique, but (2.8) does not reduce to 1 unless $\gamma_1 = \gamma_2$.

2.4 Nearest-neighbour Markov point processes

In the previous section, we encountered the continuum random cluster model with conditional intensity given by

$$\lambda(u; \mathbf{x}) = \beta \, 2^{c(\mathbf{x} \cup \{u\}) - c(\mathbf{x})} \tag{2.9}$$

where $\beta > 0$ is an intensity parameter and $c(\mathbf{z})$ denotes the number of connected components in $\bigcup_{z \in \mathbf{z}} B(z, R/2)$, the union of balls centred at the points of $\mathbf{z}$. As noted in example 2.10, $\lambda(u; \mathbf{x})$ may depend on points further than R away from u. However, any $\xi \in \mathbf{x}$ not belonging to the connected component of u in the set $\bigcup_{x \in \mathbf{x} \cup \{u\}} B(x, R/2)$ does not affect $c(\mathbf{x} \cup \{u\}) - c(\mathbf{x})$.

More formally, for each pattern $\mathbf{x}$ a symmetric, reflexive relation can be defined on the points of $\mathbf{x}$ by $u \underset{\mathbf{x}}{\sim} v$ if and only if u and v belong to the same connected component of $\bigcup_{x \in \mathbf{x}} B(x, R/2)$. Doing so, (2.9) only depends on $\{x \in \mathbf{x} : x \underset{\mathbf{x} \cup \{u\}}{\sim} u\}$ and on the connectedness relations $\underset{\mathbf{x}}{\sim}$ and $\underset{\mathbf{x} \cup \{u\}}{\sim}$ restricted to this set. Thus, (2.9) expresses an interaction structure similar to that of a Markov process (cf. definition 2.2), except for the fact that the 'neighbours' of a newly added point u depend on the current configuration $\mathbf{x}$.

The above observation underlies the notion of a *nearest-neighbour Markov point process* [11]. Before giving the definition, we need to specify a family of configuration-dependent neighbourhood relations [11].

Definition 2.5 Let $\underset{\mathbf{x}}{\sim}$ be a symmetric, reflexive relation on $\mathbf{x} \in N^f$. The $\mathbf{x}$–neighbourhood of a subset $\mathbf{z} \subseteq \mathbf{x}$ is defined as

$$\partial(\mathbf{z} \mid \mathbf{x}) = \{x \in \mathbf{x} : x \underset{\mathbf{x}}{\sim} z \text{ for some } z \in \mathbf{z}\}.$$

The configuration $\mathbf{z}$ is an $\mathbf{x}$-clique if for each $u, v \in \mathbf{z}$, $u \underset{\mathbf{x}}{\sim} v$.

Suppose for each configuration $\mathbf{x}$ a neighbourhood relation $\underset{\mathbf{x}}{\sim}$ is defined on $\mathbf{x}$, and we want to build a pairwise interaction model by setting

$$p(\mathbf{x}) = \alpha \prod_{u \underset{\mathbf{x}}{\sim} v} \gamma(u, v)$$

for some Borel measurable interaction function $\gamma(\cdot,\cdot)$ that is symmetric in its arguments. Then

$$\frac{p(\mathbf{x} \cup \{u\})}{p(\mathbf{x})} = \prod_{\substack{x \underset{\mathbf{x}\cup\{u\}}{\sim} u}} \gamma(x,u) \frac{\prod_{\substack{x \underset{\mathbf{x}\cup\{u\}}{\sim} y}} \gamma(x,y)}{\prod_{x \underset{\mathbf{x}}{\sim} y} \gamma(x,y)}. \tag{2.10}$$

It is clear from this expression that in order to obtain a realisation-dependent Markov property for $p(\cdot)$, consistency conditions [11] must be imposed on the family of neighbourhood relations.

Definition 2.6 Let $(\mathcal{X}, d)$ be a complete, separable metric space, and $\mathbf{y} \subseteq \mathbf{z} \in N^{\mathrm{f}}$, $u, v \in \mathcal{X}$ such that $u, v \notin \mathbf{z}$. Define consistency conditions as follows.

(C1) $\chi(\mathbf{y} \mid \mathbf{z}) \neq \chi(\mathbf{y} \mid \mathbf{z} \cup \{u\})$ implies $\mathbf{y} \subseteq \partial(\{u\}\| \mid \mathbf{z} \cup \{u\})$;

(C2) if $u \underset{\mathbf{x}}{\not\sim} v$ where $\mathbf{x} = \mathbf{z} \cup \{u,v\}$ then

$$\chi(\mathbf{y} \mid \mathbf{z} \cup \{u\}) + \chi(\mathbf{y} \mid \mathbf{z} \cup \{v\}) = \chi(\mathbf{y} \mid \mathbf{z}) + \chi(\mathbf{y} \mid \mathbf{x}).$$

where χ is the clique indicator function, i.e. $\chi(\mathbf{y} \mid \mathbf{x}) = \mathbb{I}\{\mathbf{y}$ is an $\mathbf{x}$-clique $\}$.

Clearly, condition (C1) is needed to ensure that (2.10) is a function of the 'neighbours' of u; (C2) is needed to allow a Hammersley–Clifford style factorisation theorem [11].

Example 2.15 Let $\sim$ be a symmetric, reflexive relation on $\mathcal{X}$ and define

$$u \underset{\mathbf{x}}{\sim} v \Leftrightarrow u \sim v$$

for any $\mathbf{x} \in N^{\mathrm{f}}$. Since $\underset{\mathbf{x}}{\sim}$ does not depend on the configuration $\mathbf{x}$, evidently conditions (C1) and (C2) hold.

Example 2.16 Given a symmetric, reflexive relation $\sim$ on $\mathcal{X}$, [11] define the family of *connected component relations* $\{\underset{\mathbf{x}}{\sim} : \mathbf{x} \in N^{\mathrm{f}}\}$ by $u \underset{\mathbf{x}}{\sim} v$ if u and v are connected by a path in $\mathbf{x}$, that is $u \sim z_1 \sim \cdots \sim z_n \sim v$ for some $z_1, \ldots, z_n \in \mathbf{x}$ $(n \geq 0)$. Clearly, $\underset{\mathbf{x}}{\sim}$ depends on the configuration $\mathbf{x}$. For instance, consider Figure 2.4 and suppose that only points joined by a

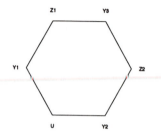

Fig. 2.4   Configuration violating consistency condition (C1).

line are $\sim$-neighbours. Then the ring is a clique in $\mathbf{x} = \{u, y_1, y_2, y_3, z_1, z_2\}$ but if z_1 and z_2 were deleted, y_3 is no longer connected to the remaining points.

To check the consistency condition (C1), first observe that if $\chi(\mathbf{y} \mid \mathbf{z}) = 1$ than a fortiori $\chi(\mathbf{y} \mid \mathbf{z} \cup \{u\}) = 1$. Thus, without loss of generality, assume $\chi(\mathbf{y} \mid \mathbf{z}) = 0$ and $\chi(\mathbf{y} \mid \mathbf{z} \cup \{u\}) = 1$. In this case, $\mathbf{y}$ belongs entirely to a single connected component in $\mathbf{z} \cup \{u\}$, but there exist y_1, y_2 in $\mathbf{y}$ that are not connected by a path in $\mathbf{z}$. Hence the path between y_1 and y_2 in $\mathbf{z} \cup \{u\}$ must contain u. It follows that the component of $\mathbf{y}$ in $\mathbf{z} \cup \{u\}$ is that of u.

Regarding condition (C2), let $u \underset{\mathbf{x}}{\not\sim} v$ for $\mathbf{x} = \mathbf{z} \cup \{u, v\}$. If $\chi(\mathbf{y} \mid \mathbf{z}) = 1$, then $\chi(\mathbf{y} \mid \mathbf{x}) = \chi(\mathbf{y} \mid \mathbf{z} \cup \{u\}) = \chi(\mathbf{y} \mid \mathbf{z} \cup \{v\}) = 1$ and the condition holds true. If $\chi(\mathbf{y} \mid \mathbf{z}) = 0$, distinguish three cases:

- suppose $\chi(\mathbf{y} \mid \mathbf{z} \cup \{u\}) = 1$. Then $\chi(\mathbf{y} \mid \mathbf{x}) = 1$ and we will prove that $\chi(\mathbf{y} \mid \mathbf{z} \cup \{v\}) = 0$ by contradiction. Since $\chi(\mathbf{y} \mid \mathbf{z}) = 0$, there exist $y_1, y_2 \in \mathbf{y}$ such that $y_1 \underset{\mathbf{z}}{\not\sim} y_2$. If $\chi(\mathbf{y} \mid \mathbf{z} \cup \{v\}) = 1$, there is a path $y_1 \sim z_1 \sim \cdots \sim z_m \sim v \sim z_{m+1} \sim \cdots \sim z_{m+n} \sim y_2$ with $z_i \in \mathbf{z}$ $(i = 1, \ldots, m + n \in \mathbb{N}_0)$. Similarly, since $\chi(\mathbf{y} \mid \mathbf{z} \cup \{u\}) =$

1, there is a path $y_1 \sim z_1' \sim \cdots \sim z_{m'}' \sim u \sim z_{m'+1}' \sim \cdots \sim z_{m'+n'}' \sim y_2$ with $z_i' \in \mathbf{z}$. But then $u \sim z_{m'+1}' \sim \cdots \sim z_{m'+n'}' \sim y_2 \sim z_{m+n} \cdots z_{m+1} \sim v$ and hence $u \underset{\mathbf{x}}{\sim} v$. This contradicts the assumption on u and v, hence $\chi(\mathbf{y} \mid \mathbf{z} \cup \{v\}) = 0$.

- by reversing the roles of u and v, if $\chi(\mathbf{y} \mid \mathbf{z} \cup \{v\}) = 1$ then $\chi(\mathbf{y} \mid \mathbf{z} \cup \{u\}) = 0$ and $\chi(\mathbf{y} \mid \mathbf{x}) = 1$, hence (C2) holds.

- it remains to check the case that $\chi(\mathbf{y} \mid \mathbf{z} \cup \{u\}) = 0$ and $\chi(\mathbf{y} \mid \mathbf{z} \cup \{v\}) = 0$. We need to prove that $\chi(\mathbf{y} \mid \mathbf{x}) = 0$. Proceed as before by contradiction, supposing that $\chi(\mathbf{y} \mid \mathbf{x}) = 1$. Since $\chi(\mathbf{y} \mid \mathbf{z} \cup \{u\}) = 0$ but $\chi(\mathbf{y} \mid \mathbf{x}) = 1$, there are $y_1, y_2 \in \mathbf{y}$ that are path-connected via v. Since $\chi(\mathbf{y} \mid \mathbf{z} \cup \{v\}) = 0$ but $\chi(\mathbf{y} \mid \mathbf{x}) = 1$, there are $y_3, y_4 \in \mathbf{y}$ that are path-connected via u. As $\chi(\mathbf{y} \mid \mathbf{x}) = 1$, y_4 is path-connected in $\mathbf{x}$ to y_1, hence $u \underset{\mathbf{x}}{\sim} v$. By assumption, $u \underset{\mathbf{x}}{\nsim} v$, hence $\chi(\mathbf{y} \mid \mathbf{x}) = 0$.

Not every family of neighbourhood relations satisfies the consistency conditions of definition 2.6.

Example 2.17 Let $\sim$ be any symmetric relation on $\mathcal{X}$. Define a relation $\underset{\mathbf{x}}{\sim}^2$ on $\mathbf{x}$ by $u \underset{\mathbf{x}}{\sim}^2 v$ if **either** u is a $\sim$–neighbour of v ($u \sim v$) **or** the $\sim$–neighbour of a neighbour of v (there exists a $w \in \mathbf{x}$ such that $u \sim w \sim v$).

Assuming we can find points $z_1, z_2 \in \mathbf{z} \setminus \mathbf{y}$, $y_1, y_2, y_3 \in \mathbf{y}$ such that

$$y_1 \sim z_1 \sim y_3 \sim z_2 \sim y_2 \sim u \sim y_1$$

are the only neighbours in $\mathbf{z} \cup \{u\}$ (refer to Figure 2.4), it is readily seen that $\chi(\mathbf{y} \mid \mathbf{z}) = 0$, $\chi(\mathbf{y} \mid \mathbf{z} \cup \{u\}) = 1$, but $y_3 \notin \partial(\{u\} \mid \mathbf{z} \cup \{u\})$. Hence (C1) fails.

We are now ready to state the main definition of this section [11].

Definition 2.7 Let $(\mathcal{X}, d)$ be a complete, separable metric space, $\nu(\cdot)$ a finite, non-atomic Borel measure, and $\pi_\nu(\cdot)$ the distribution of a Poisson process on $\mathcal{X}$ with intensity measure $\nu(\cdot)$. Let $\{\underset{\mathbf{x}}{\sim} : \mathbf{x} \in N^{\mathrm{f}}\}$ be a family of reflexive, symmetric relations on $\mathbf{x} \in N^{\mathrm{f}}$.

Let X be a point process on $\mathcal{X}$ specified by means of a density $p(\cdot)$ with respect to $\pi_\nu(\cdot)$. Then X is a *nearest-neighbour Markov point process* with respect to $\{\underset{\mathbf{x}}{\sim} : \mathbf{x} \in N^f\}$ if for all $\mathbf{x} \in N^f$ such that $p(\mathbf{x}) > 0$,

(a) $p(\mathbf{y}) > 0$ for all $\mathbf{y} \subseteq \mathbf{x}$;

(b) for all $u \in \mathcal{X}$, $p(\mathbf{x} \cup \{u\})/p(\mathbf{x})$ depends only on u, on $\partial(\{u\} \mid \mathbf{x} \cup \{u\}) \cap$
$\mathbf{x} = \{x_i \in \mathbf{x} : x_i \underset{\mathbf{x} \cup \{u\}}{\sim} u\}$ and on the relations $\underset{\mathbf{x}}{\sim}$, $\underset{\mathbf{x} \cup \{u\}}{\sim}$ restricted
to $\partial(\{u\} \mid \mathbf{x} \cup \{u\}) \cap \mathbf{x}$.

Example 2.18   Consider the continuum random cluster process X on a
compact subset $\mathcal{X}$ of $\mathbb{R}^2$ with positive volume $\mu(\mathcal{X}) < \infty$. Its density with
respect to a unit rate Poisson process on $\mathcal{X}$ is given by [107; 143]

$$p(\mathbf{x}) = \alpha\, e^{-\mu(\mathcal{X})}\, \beta^{n(\mathbf{x})}\, 2^{c(\mathbf{x})}$$

where $c(\mathbf{x})$ denotes the number of connected components in $\cup_{x \in \mathbf{x}} B(x, R/2)$.
The conditional intensity (2.9) depends on $\mathbf{x}$ only through $c(\mathbf{x} \cup \{u\}) - c(\mathbf{x})$,
hence X is a nearest-neighbour Markov point process with respect to the
connected component relation (example 2.16) based on

$$u \sim v \Leftrightarrow \|u - v\| \leq R.$$

Example 2.19   Consider the pairwise interaction model [11]

$$p(\mathbf{x}) = \alpha\, \beta^{n(\mathbf{x})} \prod_{\underset{\mathbf{x}}{u \sim v}} \gamma(u, v)$$

where $\beta > 0$ is the intensity parameter, and $\gamma : \mathcal{X} \times \mathcal{X} \to (0, \infty)$ is a Borel
measurable, symmetric interaction function. If the family $\{\underset{\mathbf{x}}{\sim} : \mathbf{x} \in N^f\}$
satisfies (C1), $p(\cdot)$ defines a nearest-neighbour Markov point process.
Indeed, in the conditional intensity (2.10), the term

$$\prod_{\underset{\mathbf{x} \cup \{u\}}{x \sim u}} \gamma(x, u)$$

depends on u, on $\underset{\mathbf{x} \cup \{u\}}{\sim}$ and $\partial(\{u\} | \mathbf{x} \cup \{u\})$. As for the term

$$\frac{\prod_{\underset{\mathbf{x} \cup \{u\}}{x \sim y}} \gamma(x, y)}{\prod_{\underset{\mathbf{x}}{x \sim y}} \gamma(x, y)},$$

note that condition (C1) with $\mathbf{z} = \mathbf{x}$ and $\mathbf{y} = \{x, y\}$ implies that whenever $\chi(\mathbf{y} \mid \mathbf{x}) \neq \chi(\mathbf{y} \mid \mathbf{x} \cup \{u\})$, then $x, y \in \partial(\{u\} | \mathbf{x} \cup \{u\})$. Therefore (2.10) depends only on $\partial(\{u\} | \mathbf{x} \cup \{u\}) \cap \mathbf{x}$ and the relations $\underset{\mathbf{x}}{\sim}$, $\underset{\mathbf{x} \cup \{u\}}{\sim}$ restricted to this neighbourhood.

Example 2.20 Consider two independent Poisson processes X_1, X_2 on an interval $(a, b) \subset \mathbb{R}$, conditioned to respect a hard core distance $R > 0$ between the points of different components [221]. Then the marginal distribution of X_1 has density

$$p(\mathbf{x}_1) = \alpha \beta^{n(\mathbf{x}_1)} \exp[-\beta \operatorname{length}(U_{\mathbf{x}_1})]$$

with respect to a unit rate Poisson process, where $U_{\mathbf{x}_1} = \bigcup_{x \in \mathbf{x}_1} [x - R, x + R]$ denotes the union of intervals around points in $\mathbf{x}_1$. Writing $\mathbf{x}_1 = \{x_1, \ldots, x_n\}$ with $x_1 < x_2 < \cdots < x_n$, we obtain

$$p(\mathbf{x}_1) = \alpha \beta^n \exp\left[-2\beta R - \beta \sum_{i=1}^{n-1} \min(x_{i+1} - x_i, 2R)\right],$$

a Markov density with respect to the nearest-neighbour relation [11; 221]

$$x_{i-1} \underset{\mathbf{x}}{\sim} x_i \underset{\mathbf{x}}{\sim} x_{i+1}$$

(with obvious modification for $i = 0, n$) in which the $\mathbf{x}$-neighbours of a point are their immediate predecessor and successor in $\mathbf{x}$, regardless of how far away they are.

For a family of symmetric relations $\{\underset{\mathbf{x}}{\sim} : \mathbf{x} \in N^f\}$ satisfying conditions (C1) and (C2), the Hammersley–Clifford theorem states that $p(\cdot)$ is a Markov density if and only if

$$p(\mathbf{x}) = \prod_{\mathbf{y} \subseteq \mathbf{x}} \phi(\mathbf{y})^{\chi(\mathbf{y} \mid \mathbf{x})} \tag{2.11}$$

(taking $0^0 = 0$) for some measurable interaction function $\phi : N^f \to [0, \infty)$ that is hereditary in the sense that $\phi(\mathbf{x}) > 0$ implies (a) $\phi(\mathbf{y}) > 0$ for all $\mathbf{y} \subseteq \mathbf{x}$ and (b) if moreover $\phi(\{u\} \cup \partial(\{u\} \mid \mathbf{x} \cup \{u\})) > 0$ then $\phi(\mathbf{x} \cup \{u\}) > 0$.

Rather than prove this result in general [11], we will restrict ourselves to the important special case of the connected component relation [10], see section 2.5 below.

2.5   Connected component Markov point processes

As for Ripley–Kelly Markov point processes (section 2.1), under certain regularity conditions, a spatial Markov property [102] and Hammersley–Clifford factorisation [11] hold for nearest-neighbour models too. However, the expressions become much more complicated, cf. (2.11). Thus, for concreteness' sake, we will focus on the connected components relation introduced in example 2.16.

The next theorem gives a Hammersley–Clifford style factorisation [10; 143] for connected component Markov point processes.

Theorem 2.3   *A point process with density* $p(\cdot)$ *is nearest–neighbour Markov with respect to the connected components relation if and only if it can be factorised as*

$$p(\mathbf{x}) = \prod_{\underset{\mathbf{x}}{\sim}\text{-cliques } \mathbf{y}} \phi(\mathbf{y}) \tag{2.12}$$

where $\phi : N^{\mathrm{f}} \to [0, \infty)$ *is a measurable fuction such that if* $\mathbf{y}$ *is a* $\underset{\mathbf{y}}{\sim}$-*clique with* $\phi(\mathbf{y}) > 0$, *then* $\phi(\mathbf{z}) > 0$ *for all configurations* $\mathbf{z} \subseteq \mathbf{y}$.

Formula (2.12) should be compared to (2.7).

Proof.   Suppose that $p(\cdot)$ is of the form (2.12). Let $\mathbf{x} \in N^{\mathrm{f}}$, $u \in \mathcal{X}$ and write $\mathbf{x}_1, \dots, \mathbf{x}_k$, $\mathbf{w} \cup \{\mathbf{u}\}$ for the connected components of $\mathbf{x} \cup \{u\}$. Then, if $\mathbf{x}_{k+1}, \dots, \mathbf{x}_l$ are the connected components of $\mathbf{w}$, the connected components of the configuration $\mathbf{x}$ are $\mathbf{x}_1', \dots, \mathbf{x}_l$. Since $\mathbf{x} \cup \{u\}$-cliques are subsets of the connected components of $\mathbf{x} \cup \{u\}$,

$$p(\mathbf{x} \cup \{u\}) = \phi(\emptyset) \left[\prod_{i=1}^{k} \prod_{\emptyset \neq \mathbf{y} \subseteq \mathbf{x}_i} \phi(\mathbf{y}) \right] \prod_{\emptyset \neq \mathbf{y} \subseteq \mathbf{w} \cup \{\mathbf{u}\}} \phi(\mathbf{y})$$

and similarly

$$p(\mathbf{x}) = \phi(\emptyset) \left[\prod_{i=1}^{k} \prod_{\emptyset \neq \mathbf{y} \subseteq \mathbf{x}_i} \phi(\mathbf{y}) \right] \left[\prod_{i=k+1}^{l} \prod_{\emptyset \neq \mathbf{y} \subseteq \mathbf{x}_i} \phi(\mathbf{y}) \right].$$

Hence $p(\mathbf{x} \cup \{u\}) > 0$ implies $p(\mathbf{x}) > 0$. Moreover, the conditional intensity

$$\lambda(u; \mathbf{x}) = \frac{\prod_{\emptyset \neq \mathbf{y} \subseteq \mathbf{w} \cup \{u\}} \phi(\mathbf{y})}{\prod_{i=k+1}^{l} \prod_{\emptyset \neq \mathbf{y} \subseteq \mathbf{x}_i} \phi(\mathbf{y})}$$

depends only on the added point u, its neighbourhood $\mathbf{n} = \partial(\{u\} \mid \mathbf{x} \cup \{u\}) \cap \mathbf{x}$ and the relations $\underset{\mathbf{x}}{\sim}$, $\underset{\mathbf{x} \cup \{u\}}{\sim}$ restricted to $\mathbf{n}$. It follows that X is a connected component Markov point process.

For the converse, suppose that X is a connected component Markov point process. As in the proof of the Hammersley–Clifford theorem for Ripley–Kelly Markov point processes, inductively define an interaction function $\phi : N^f \to [0, \infty)$ by

$$\phi(\emptyset) = p(\emptyset)$$

$$\phi(\mathbf{x}) = \frac{p(\mathbf{x})}{\prod_{\mathbf{x} - \text{cliques}\, \mathbf{y}: \mathbf{x} \neq \mathbf{y} \subset \mathbf{x}} \phi(\mathbf{y})}$$

with the convention $0/0 = 0$. Note that if $\prod_{\mathbf{x} - \text{cliques}\, \mathbf{y}: \mathbf{x} \neq \mathbf{y} \subset \mathbf{x}} \phi(\mathbf{y}) = 0$, necessarily $p(\mathbf{y})$ must be zero for some proper subset $\mathbf{y} \subset \mathbf{x}$, and by property (a) in definition 2.7, $p(\mathbf{x}) = 0$. Hence $\phi(\cdot)$ is well-defined.

It is clear that whenever the clique indicator $\chi(\mathbf{y} \mid \mathbf{y}) = 1$, any subset $\mathbf{z} \subseteq \mathbf{y}$ is connected in $\mathbf{y}$. Therefore, if $\phi(\mathbf{y}) > 0$, $\phi(\mathbf{z})$ is strictly positive as well.

Finally, to prove the factorisation, we use induction on the number of points. By definition the factorisation holds for the empty set. Assume the factorisation holds for configurations with up to $n - 1$ points, and consider a pattern $\mathbf{x}$ of cardinality $n \geq 1$. First note that if $p(\mathbf{x}) = 0$, the factorisation is immediate. Thus, suppose that $p(\mathbf{x}) > 0$. Since X is a connected component Markov point process, $p(\mathbf{y})$ and hence $\phi(\mathbf{y})$ are strictly positive for all $\mathbf{x} \neq \mathbf{y} \subset \mathbf{x}$. Consequently, if $\chi(\mathbf{x} \mid \mathbf{x}) = 1$, $p(\mathbf{x}) = \prod_{\mathbf{y} \subset \mathbf{x}} \phi(\mathbf{y})$ by the definition of $\phi(\mathbf{x})$. If $\chi(\mathbf{x} \mid \mathbf{x}) = 0$, write $\mathbf{x} = \mathbf{z} \cup \{u, v\}$ for some $u, v \notin \mathbf{z}$ such that $u \underset{\mathbf{x}}{\not\sim} v$. Since $p(\mathbf{x}) > 0$, by (a) in definition 2.7, both $p(\mathbf{z})$ and

$p(z \cup \{u\})$ are strictly positive and we may write

$$p(\mathbf{x}) = \frac{p(z \cup \{u, v\})}{p(z \cup \{u\})} \, p(z \cup \{u\}) = \frac{p(z \cup \{v\})}{p(z)} \, p(z \cup \{u\})$$

using the local Markov property for the last equation. By the induction hypothesis,

$$p(\mathbf{x}) = \frac{\prod_{z \cup \{v\} - \text{cliques } \mathbf{y}} \phi(\mathbf{y})}{\prod_{z - \text{cliques } \mathbf{y}} \phi(\mathbf{y})} \prod_{z \cup \{u\} - \text{cliques } \mathbf{y}} \phi(\mathbf{y}).$$

The proof is complete if we observe that, since $u \underset{\mathbf{x}}{\not\sim} v$, an $\mathbf{x}$-clique is either a $z \cup \{u\}$ clique or a $z \cup \{v\}$-clique that is not connected in z.   □

The connected components relation has the nice property that if a configuration $\mathbf{y}$ is an $\mathbf{x}$-clique, then $\mathbf{y}$ is a clique in any $\mathbf{x}' \supseteq \mathbf{x}$. This is not the case for a general family of relations $\{\underset{\mathbf{x}}{\sim} : \mathbf{x} \in N^f\}$, and the interaction function $\phi(\mathbf{y})$ might be zero for a pattern $\mathbf{y}$ that is an $\mathbf{x}$-clique but not a clique in $\mathbf{x}'$. As a result, in (2.11) the clique indicator function has to be taken into account explicitly.

In order to derive a spatial Markov property [143] for connected component Markov point processes, note that (2.12) is equivalent to

$$p(\mathbf{x}) = p(\emptyset) \prod_{i=1}^{k} \Phi(\mathbf{x}_i) \tag{2.13}$$

where $\mathbf{x}_i$, $(i = 1, \ldots, k)$ are the (maximal) connected components of $\mathbf{x}$, and $\Phi(\mathbf{x}_i) = \prod_{\emptyset \neq \mathbf{y} \subseteq \mathbf{x}_i} \phi(\mathbf{y})$. The *component interaction function* $\Phi(\cdot)$ is hereditary in the sense that if $\chi(\mathbf{x} \mid \mathbf{x}) = 1$, and $\mathbf{y} \subseteq \mathbf{x}$ with $\chi(\mathbf{y} \mid \mathbf{y}) = 1$, then $\Phi(\mathbf{x}) > 0$ implies $\Phi(\mathbf{y}) > 0$.

Let us suppose that the point process X is observed in some subwindow $A \subseteq \mathcal{X}$ and let $\mathbf{x}_i$, $i = 1, \ldots, l$, be those connected components of $\mathbf{x} \cap A$ for which the neighbourhood $\partial(\mathbf{x}_i) \cap A^c$ with respect to the underlying relation $\sim$ is non-empty. In other words, $\mathbf{x}_i$ are the components that may contain a neighbour of a point in A^c. For the special case $u \sim v \Leftrightarrow \|u - v\| \leq R$, these components are represented by the centers of the discs in Figure 2.5. Let I be the 'interior' set, as illustrated in Figure 2.5. Then it follows from the

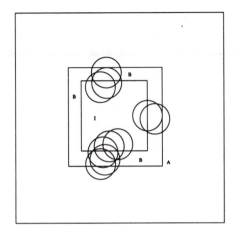

Fig. 2.5 For the connected components relation at range R, the boundary set B consists of the window A intersected with the set of R-neighbours of A^c and R-neighbours of the components intersecting the boundary of A. The interior set is the complement of the boundary set, $I = A \setminus B$.

factorisation (2.13) that the conditional distribution of $X \cap I$ given $X \cap I^c$ depends only on $\mathbf{x}_i$, $i = 1, \ldots, l$.

Example 2.21 For the continuum random cluster model (cf. example 2.10) the component interaction function is

$$\Phi(\mathbf{x}) = 2\,\beta^{n(\mathbf{x})}$$

where $n(\mathbf{x})$ is the cardinality of $\mathbf{x}$.

We conclude this section with a few examples of connected component *marked* point processes.

Example 2.22 Let $\mathcal{X}$ be a compact window in $\mathbb{R}^2$, and consider the marked point process on $\mathcal{X} \times \{1, \ldots, M\}$ defined by

$$p(\mathbf{y}) = \alpha \prod_{(x,k) \in \mathbf{y}} \beta_k \prod_{(u,k) \underset{\mathbf{y}}{\sim} (v,l)} \gamma_{kl}(\|u - v\|)$$

for some $\alpha > 0$, intensity parameters $\beta_k > 0$, and Borel measurable interaction functions $\gamma_{kl} : [0, \infty) \to [0, \infty)$. The reference distribution is that of a unit rate Poisson process of locations marked by a uniformly distributed label. As always, for any particular choice of $\gamma_{kl}(||u - v||)$, one must verify that the resulting density $p(\cdot)$ is integrable. For instance, if

$$(u, k) \sim (v, l) \Leftrightarrow ||u - v|| \leq r_{kl}$$

and $\gamma_{kl}(r) \equiv \gamma_{kl} = \gamma_{lk}$ for some $\gamma_{kl} \in (0, 1)$, $p(\cdot)$ is well-defined and discourages configurations with many large components [11].

Example 2.23   Consider the bivariate pairwise interaction model of example 2.12.   We derived the marginal probability density of the first component as

$$p(\mathbf{x}_1) = \alpha \beta_1^{n(\mathbf{x}_1)} \prod_{u,v \in \mathbf{x}_1} \gamma_{11}(||u - v||) I_2(\mathbf{x}_1)$$

where $n(\cdot)$ denotes cardinality and

$$I_2(\mathbf{x}_1) = \int_{N^f} \beta_2^{n(\mathbf{x}_2)} \prod_{u,v \in \mathbf{x}_2} \gamma_{22}(||u - v||) \prod_{u \in \mathbf{x}_1, v \in \mathbf{x}_2} \gamma_{12}(||u - v||) \, d\pi(\mathbf{x}_2)$$

is an integral with respect to the distribution of the reference Poisson process. Now, if $\gamma_{22} \equiv 1$ so there are no interactions within the second component, then

$$I_2(\mathbf{x}_1) = \int \beta^{n(\mathbf{x}_2)} \prod_{u \in \mathbf{x}_1, v \in \mathbf{x}_2} \gamma_{12}(||u - v||) \, d\pi(\mathbf{x}_2) = I_2(\emptyset) \prod_{i=1}^{k} \Phi(\mathbf{x}_{1i})$$

where

$$\Phi(\mathbf{y}) = \mathbf{E}_{\beta,\mathbf{y}} \left[\prod_{u \in \mathbf{y}, v \in X} \gamma_{12}(||u - v||) \right]$$

where $\mathbf{x}_{1i}$, $i = 1, \ldots, k$ are the connected components of $\mathbf{x}_1$ at range $2r = 2r_{12}$, $\mathbf{E}_{\beta,\mathbf{y}}$ denotes the expectation under a Poisson process of rate β on $\mathbf{y}^{(r)}$, the r-envelope of $\mathbf{y}$ in $\mathcal{X}$.   Hence X_1 is a connected component Markov point process.

The last lemma of this chapter states that the superposition of two connected component point processes is again connected component Markov [34].

A fortiori, the superposition of two Ripley–Kelly Markov point processes is a connected component Markov point process.

Lemma 2.2 *Let X_1 and X_2 be independent and identically distributed connected component Markov point processes on a complete, separable metric space $(\mathcal{X}, d)$, defined on a common underlying probability space. Let $X = X_1 \cup X_2$ be the superposition of X_1 and X_2. Then, if X_1 has density $p(\cdot)$ with respect to a Poisson process on $\mathcal{X}$ with finite, non-atomic intensity measure $\nu(\cdot)$, X is a connected component Markov point process.*

Proof. As in lemma 2.1,

$$p_s(\mathbf{x}) = e^{-\nu(\mathcal{X})} \sum_{\mathbf{y},\mathbf{z}} p(\mathbf{y}) \, p(\mathbf{z})$$

where the sum is over all ordered partitions of $\mathbf{x}$ into two groups $\mathbf{y}$ and $\mathbf{z}$, and $p_s(\cdot)$ is hereditary. By (2.13), $p_s(\mathbf{x})$ can be written as

$$e^{-\nu(\mathcal{X})} p(\emptyset)^2 \sum_{\mathbf{y},\mathbf{z}} \left[\prod_{i=1}^{c(\mathbf{y})} \Phi(\mathbf{y}_i) \prod_{i=1}^{c(\mathbf{z})} \Phi(\mathbf{z}_i) \right] \qquad (2.14)$$

where $\mathbf{y}_i$, $i = 1, \ldots, c(\mathbf{y})$ are the connected components in $\mathbf{y}$, while the connected components in $\mathbf{z}$ are $\mathbf{z}_i$, $i = 1, \ldots, c(\mathbf{z})$. Let $\mathbf{x}_1, \ldots, \mathbf{x}_k$ be the connected components in the unmarked pattern $\mathbf{x}$. Since any connected component of $\mathbf{y}$ or $\mathbf{z}$ must be a subset of one of the $\mathbf{x}_i$s, (2.14) reduces to

$$e^{-\nu(\mathcal{X})} p(\emptyset)^2 \left[\sum_{\mathbf{y}_1, \mathbf{z}_1 \subseteq \mathbf{x}_1} \left(\prod_{i=1}^{c(\mathbf{y}_1)} \Phi(\mathbf{y}_{1i}) \prod_{i=1}^{c(\mathbf{z}_1)} \Phi(\mathbf{z}_{1i}) \right) \right] \cdots$$

$$\left[\sum_{\mathbf{y}_k, \mathbf{z}_k \subseteq \mathbf{x}_k} \left(\prod_{i=1}^{c(\mathbf{y}_k)} \Phi(\mathbf{y}_{ki}) \prod_{i=1}^{c(\mathbf{z}_k)} \Phi(\mathbf{z}_{ki}) \right) \right]$$

where $\mathbf{y}_{ji}, \mathbf{z}_{ji}$ are the connected components of the partition $(\mathbf{y}_j, \mathbf{z}_j)$ of $\mathbf{x}_j$. Hence $p_s(\mathbf{x})$ is of the form (2.13) and X is a connected component Markov point process based on $\sim$. $\qquad \square$

Chapter 3

Statistical Inference

3.1 Introduction

In section 2.1, we met a molecular biologist searching for a probabilistic model to explain cell patterns observed under a microscope. Having postulated a hard core process, he naturally wants to estimate the parameters, assess the goodness of fit, and compute the average values of interesting quantities under the fitted model. However, he soon finds himself bogged down by the intractable normalising constant.

In these circumstances, the *Monte Carlo approach* provides an alternative way to proceed. Named after the casino in the capital of Monaco, Monte Carlo methods comprise that branch of mathematics concerned with random samples as a computational tool. It should be noted that such methods are not confined to the realm of stochastic processes; Monte Carlo techniques have been applied successfully to the evaluation of intractable integrals, to solving sets of operator equations, as well as to a wide range of optimisation problems [1; 22; 23; 25; 56; 66; 80; 198; 203; 204; 218].

Unfortunately, for point processes the same factors that hamper computation also complicate simulation: the sample space N^f is complex, and moreover the density as a rule involves a normalising constant that is not available in closed form. On the other hand, the conditional intensity of a Markov point process has a simple local form. In a pioneering paper, Metropolis et al. [137] proposed to run a Markov process in time with the target distribution as its limit and with transitions governed by the conditional intensity. If the process is run for a sufficiently long time, an approximate sample from the equilibrium distribution is obtained. Of course, the particular Markov process proposed in [137] (cf. section 3.2) is not the only one with the required equilibrium. There are many such processes, the art is to find a transition kernel that is easy to implement and mixes rapidly. In sections 3.2–3.4, we will describe some generally applicable strategies for the design of Markov chain Monte Carlo (MCMC) algorithms and give conditions for convergence to the target distribution.

Once a Monte Carlo sampler has been defined, the practitioner is faced with the problem of when to stop. Although some theory on the rate of convergence has been developed, it is usually of little practical help. Sometimes however, as we will see in section 3.5, it is possible to let the algorithm itself indicate when equilibrium is reached, so that exact rather than approximate samples [103; 105; 167] are obtained.

Except for generating samples from a given point process distribution, MCMC ideas are very helpful to tackle the classical inference tasks of parameter estimation and hypothesis testing. In section 3.7, will discuss how (approximate) maximum likelihood estimators can be obtained [63; 64; 143]. Computationally less demanding alternatives are described in sections 3.8.

3.2   The Metropolis–Hastings algorithm

Throughout this section, suppose a Markov point process on a complete, separable metric space $(\mathcal{X}, d)$ is defined by its density $p(\cdot)$ with respect to a Poisson process with finite, non-atomic intensity measure $\nu(\cdot)$.

One way of constructing a Markov process with limit distribution defined by $p(\cdot)$ is the *Metropolis–Hastings algorithm*, originally introduced in statistical physics [14; 137] to compute the behaviour of a given number of interacting molecules in a box. In the 1970s, the method received some interest in the statistical community [86; 163; 154], but it became widely known only a decade later following the successful application of Markov chain Monte Carlo, or MCMC, methods in image analysis (see eg. the papers collected in [126]). Since then, MCMC has been highly influential in the development of modern Bayesian statistics [66].

Metropolis–Hastings algorithms are discrete time Markov processes X_0, $X_1, \ldots$ where the transitions are defined in two steps: a proposal for a new state is made that is subsequently accepted or rejected based on the likelihood of the proposed state compared to the old one. The proposals must be easy to perform; a common choice [63; 64; 71; 143] is to consider changing the current configuration at one point only by addition (*birth*) or deletion (*death*). Note that for these transitions the likelihood ratio of the new state compared to the old one is the (reciprocal) conditional intensity.

To be more specific, let us assume that at time $n \geq 0$ the process is in state $\mathbf{x}$. There are two options for the proposal – a birth or a death – with respective probabilities $q(\mathbf{x})$ and $1 - q(\mathbf{x})$ depending on $\mathbf{x}$. For a birth, a new point u is sampled from a probability density $b(\mathbf{x}, \cdot)$ with respect to $\nu(\cdot)$ on $\mathcal{X}$. The proposed new state $X_{n+1} = \mathbf{x} \cup \{u\}$ is accepted with probability $A(\mathbf{x}, \mathbf{x} \cup \{u\})$, otherwise the state remains unchanged, that is $X_{n+1} = \mathbf{x}$. Similarly in case of a death, the point x to be deleted is picked according to a discrete probability distribution $d(\mathbf{x}, \cdot)$ on $\mathbf{x}$, and, if the proposal $X_{n+1} = \mathbf{x} \setminus \{x\}$ is rejected (say with probability $1 - A(\mathbf{x}, \mathbf{x} \setminus \{x\})$), the Markov process stays put at $X_{n+1} = \mathbf{x}$.

Summarising, the Metropolis–Hastings procedure for point patterns is as follows [63; 64; 71; 143].

Algorithm 3.1 Initialise $X_0 = \mathbf{x}_0$ for some configuration $\mathbf{x}_0 \in N^{\mathrm{f}}$ with $p(\mathbf{x}_0) > 0$. For $n = 0, 1, \ldots$, if $X_n = \mathbf{x}$,

- with probability $q(\mathbf{x})$, propose a *birth*: sample u from $b(\mathbf{x}, u)$ and accept the transition $X_{n+1} = \mathbf{x} \cup \{u\}$ with probability $A(\mathbf{x}, \mathbf{x} \cup \{u\})$;
- with probability $1 - q(\mathbf{x})$, propose a *death*: with probability $d(\mathbf{x}, x)$ consider deleting point $x \in \mathbf{x}$ and accept the transition $X_{n+1} = \mathbf{x} \setminus \{x\}$ with probability $A(\mathbf{x}, \mathbf{x} \setminus \{x\})$ (if $\mathbf{x} = \emptyset$ do nothing).

If the transition is rejected, the new state will be $X_{n+1} = \mathbf{x}$.

The transition kernel $Q(F \mid \mathbf{x}) = \mathbf{P}(X_{n+1} \in F \mid X_n = \mathbf{x})$ for the probability of a transition from $\mathbf{x}$ into a set $F \in \mathcal{N}^{\mathrm{f}}$ is

$$
\begin{aligned}
Q(F \mid \mathbf{x}) \; = \;\; & q(\mathbf{x}) \int_{\mathcal{X}} b(\mathbf{x}, u) \, A(\mathbf{x}, \mathbf{x} \cup \{u\}) \; \mathbb{I}\{\mathbf{x} \cup \{u\} \in F\} \, d\nu(u) \\
+ \;\; & (1 - q(\mathbf{x})) \sum_{x \in \mathbf{x}} d(\mathbf{x}, x) \, A(\mathbf{x}, \mathbf{x} \setminus \{x\}) \; \mathbb{I}\{\mathbf{x} \setminus \{x\} \in F\} \\
+ \;\; & \mathbb{I}\{\mathbf{x} \in F\} \left[q(\mathbf{x}) \int_{\mathcal{X}} b(\mathbf{x}, u) \, (1 - A(\mathbf{x}, \mathbf{x} \cup \{u\})) \, d\nu(u) \right. \\
+ \;\; & \left. (1 - q(\mathbf{x})) \sum_{x \in \mathbf{x}} d(\mathbf{x}, x) \, (1 - A(\mathbf{x}, \mathbf{x} \setminus \{x\})) \right]
\end{aligned}
$$

bearing in mind that when $\mathbf{x} = \emptyset$, the two terms involving deaths are replaced by $(1 - q(\emptyset)) \, \mathbb{I}\{\emptyset \in F\}$.

The general framework of algorithm 3.1 leaves us with plenty of choices for $b(\cdot, \cdot)$, $d(\cdot, \cdot)$ and $q(\cdot)$. However, since the process is intended to have equilibrium distribution $p(\cdot)$, a natural condition is that of *detailed balance* (and hence time-reversibility) of the Markov process:

$$
q(\mathbf{x}) \, b(\mathbf{x}, u) \, A(\mathbf{x}, \mathbf{x} \cup \{u\}) \, p(\mathbf{x}) =
$$

$$
(1 - q(\mathbf{x} \cup \{u\})) \, d(\mathbf{x} \cup \{u\}, u) \, A(\mathbf{x} \cup \{u\}, \mathbf{x}) \, p(\mathbf{x} \cup \{u\}) \tag{3.1}
$$

provided $p(\mathbf{x} \cup \{u\}) > 0$. In words, (3.1) states that once equilibrium is reached any transition from $\mathbf{x}$ into $\mathbf{x} \cup \{u\}$ is matched by one from $\mathbf{x} \cup \{u\}$ into $\mathbf{x}$.

Example 3.1 Suppose births and deaths are equally likely and sampled uniformly, i.e.

$$
q \equiv \frac{1}{2}, \quad b \equiv \frac{1}{\nu(\mathcal{X})} \;\; \text{and} \;\; d(\mathbf{x}, \cdot) = \frac{1}{n(\mathbf{x})}
$$

where $n(\mathbf{x})$ denotes the number of points in $\mathbf{x}$. Then (3.1) reduces to

$$
\frac{A(\mathbf{x}, \mathbf{x} \cup \{u\})}{A(\mathbf{x} \cup \{u\}, \mathbf{x})} = \frac{\nu(\mathcal{X})}{n(\mathbf{x}) + 1} \, \frac{p(\mathbf{x} \cup \{u\})}{p(\mathbf{x})}.
$$

Hence, writing $r(\mathbf{x}, u)$ for the ratio on the right hand side, more likely configurations can be favoured by setting

$$
A(\mathbf{x}, \mathbf{x} \cup \{u\}) = \min\{1, r(\mathbf{x}, u)\} \tag{3.2}
$$

hence $A(\mathbf{x} \cup \{u\}, \mathbf{x}) = \min\{1, 1/r(\mathbf{x}, u)\}$. This is the original choice [137] made by Metropolis and his collaborators, but other ratios are possible. For instance Barker [14] suggests

$$A(\mathbf{x}, \mathbf{x} \cup \{u\}) = \frac{p(\mathbf{x} \cup \{u\}) \, \nu(\mathcal{X})}{p(\mathbf{x}) \, (n(\mathbf{x}) + 1) + p(\mathbf{x} \cup \{u\}) \, \nu(\mathcal{X})}.$$

Some simulations obtained by the Metropolis algorithm will be presented in chapter 4.

Mote generally, let $p(\mathbf{x} \cup \{u\}) > 0$ and write

$$r(\mathbf{x}, u) = \frac{1 - q(\mathbf{x} \cup \{u\})}{q(\mathbf{x})} \frac{d(\mathbf{x} \cup \{u\}, u)}{b(\mathbf{x}, u)} \frac{p(\mathbf{x} \cup \{u\})}{p(\mathbf{x})} \tag{3.3}$$

for the Metropolis ratio of acceptance probabilities in (3.1). Then $A(\mathbf{x}, \mathbf{x} \cup \{u\}) = \min\{1, r(\mathbf{x}, u)\}$ and $A(\mathbf{x} \cup \{u\}, \mathbf{x}) = \min\{1, 1/r(\mathbf{x}, u)\}$ satisfy the detailed balance equations, provided $r(\mathbf{x}, u)$ is well-defined and non-zero (e.g. if $q(\cdot)$, $b(\cdot, \cdot)$, $d(\cdot, \cdot)$ and $p(\cdot)$ are strictly positive). If $p(\mathbf{x} \cup \{u\}) = 0$, set $A(\mathbf{x}, \mathbf{x} \cup \{u\}) = 0$. Thus, once the Markov process is in the hereditary set H of configurations with positive density $p(\cdot)$, it almost surely never leaves H.

In order to prove that algorithm 3.1 converges to the correct limit distribution [64], we have to recall the concept of *total variation distance* [79]. Let $\pi(\cdot)$ and $\pi'(\cdot)$ be probability measures on a common measurable space $(\mathcal{S}, \mathcal{A})$. Then their distance in total variation is defined as the maximal difference in mass on measurable subsets $A \in \mathcal{A}$

$$\| \pi - \pi' \| = \sup_{A \in \mathcal{A}} | \pi(A) - \pi'(A) |.$$

If both $\pi(\cdot)$ and $\pi'(\cdot)$ are absolutely continuous with respect to some measure $m(\cdot)$ with Radon-Nikodym derivatives $f_\pi(\cdot)$ and $f_{\pi'}(\cdot)$,

$$\| \pi - \pi \| = \frac{1}{2} \int_{\mathcal{S}} | f_\pi(s) - f_{\pi'}(s) | \, m(ds).$$

Theorem 3.1 *Let $(\mathcal{X}, d)$ be a complete, separable metric space, $\nu(\cdot)$ a finite, non-atomic Borel measure, and $\pi_\nu(\cdot)$ the distribution of a Poisson process on $\mathcal{X}$ with intensity measure $\nu(\cdot)$. Let $p(\cdot)$ be a probability density* *with respect to $\pi_\nu(\cdot)$. Suppose that $q(\emptyset) < 1$ and assume that the following reachability conditions hold for all $x \in \mathbf{x} \in N^f$ and $u \in \mathcal{X}$.*

- *$b(\mathbf{x}, u) > 0$ whenever $p(\mathbf{x} \cup \{u\}) > 0$;*
- *$d(\mathbf{x}, x) > 0$ whenever $p(\mathbf{x}) > 0$;*
- *$q(\mathbf{x} \cup \{u\}) < 1$ and $q(\mathbf{x}) > 0$ whenever $p(\mathbf{x} \cup \{u\}) > 0$.*

Then the Metropolis–Hastings algorithm with acceptance probabilities defined by (3.3) converges in total variation to the probability distribution specified by $p(\cdot)$ for p–almost all initial configurations.

Proof. From the detailed balance equations (3.1) it follows that $p(\cdot)$ is an invariant density. Now, let $F \in \mathcal{N}^f$ have positive mass under $p(\cdot)$. Then, from any $\mathbf{x}$, the Metropolis–Hastings sampler can reach F with positive probability by successively deleting all points of $\mathbf{x}$ and adding new ones. Thus, the sampler is irreducible on H. Since $q(\emptyset) < 1$, self-transitions from the empty set into itself occur with positive probability, and the sampler is aperiodic. It follows that $p(\cdot)$ defines the unique invariant probability distribution to which the sampler converges in total variation from almost all initial configurations. □

To get rid of the null set in Theorem 3.1, Harris recurrence is needed [138; 153; 159], that is for any $\mathbf{x} \in N^f$ and any $F \in \mathcal{N}^f$ with positive mass under $p(\cdot)$,

$$\mathbf{P}(X_n \in F \text{ infinitely often } \mid X_0 = \mathbf{x}) = 1.$$

Usually in our context this condition is satisfied (by the theory of Markov processes on general state spaces, under mild conditions any irreducible Metropolis kernel is Harris-recurrent [218]), and the Metropolis–Hastings sampler converges to the target density $p(\cdot)$ for all initial distributions. See [63; 64; 143; 218] for further details.

Sometimes, the interest is not so much in obtaining samples from a point process distribution as in some functional $f(\cdot)$. For instance, the mean number of points in a region of interest, the average inter-point distance, or the expected frequency of clumps or holes of various sizes provides valuable information on the intensity and interaction structure of the underlying process X. If the Markov process $(X_i)_{i \geq 0}$ is Harris recurrent with invariant probability density $p(\cdot)$, the mean value of $f(X)$ can be estimated simply by the average of $f(X_i)$ ($i = 1, \ldots, n$) over a long run. More precisely, for any measurable function $f : N^f \to \mathbb{R}$ that is integrable in absolute value wrt $p(\cdot)$,

$$\lim_{n \to \infty} \frac{1}{n} \sum_{i=1}^{n} f(X_i) = \int_{N^f} f(\mathbf{x}) \, p(\mathbf{x}) \, d\pi_{\ni}(\mathbf{x}) \qquad \text{a.s.} \qquad (3.4)$$

Despite its obvious appeal and simplicity, treating all transitions in a uniform way as in example 3.1 may result in unacceptably low acceptance probabilities, especially for models exhibiting strong interaction. In particular, when $p(\cdot)$ is highly peaked,

$$A(\mathbf{x}, \mathbf{x} \cup \{u\}) = \frac{\nu(\mathcal{X})}{n(\mathbf{x}) + 1} \frac{p(\mathbf{x} \cup \{u\})}{p(\mathbf{x})}$$

is very small for most u.

Example 3.2 Since the conditional intensity $\lambda(u; \mathbf{x}) \, d\nu(u)$ can be interpreted as the probability of having a point at u conditionally on configuration $\mathbf{x}$ outside an infinitesimal region centred at u, a birth proposal density $b(\mathbf{x}, u)$ proportional to $\lambda(u; \mathbf{x})$ tends to generate points at more plausible locations than uniformly generated births. Of course, in order to be able to normalise $\lambda(\cdot; \mathbf{x})$ into a probability density,

$$B(\mathbf{x}) = \int_{\mathcal{X}} \frac{p(\mathbf{x} \cup \{u\})}{p(\mathbf{x})} \, d\nu(u)$$

must be finite and non-zero for all $\mathbf{x}$ with $p(\mathbf{x}) > 0$. Being an integral over birth proposals, $B(\mathbf{x})$ can be interpreted as the total birth rate from configuration $\mathbf{x}$. Since there are $n(\mathbf{x})$ candidates for deletion, one may set

$$q(\mathbf{x}) = \frac{B(\mathbf{x})}{B(\mathbf{x}) + n(\mathbf{x})}$$

for the probability of a birth proposal. Keeping the deletion probabilities uniform, the Metropolis ratio equals

$$r(\mathbf{x}, u) = \frac{B(\mathbf{x}) + n(\mathbf{x})}{B(\mathbf{x} \cup \{u\}) + n(\mathbf{x} \cup \{u\})}.$$

Note that since the birth density $b(\cdot)$ involves the normalising constant $B(\mathbf{x})$, the algorithm will be slower to implement than uniform births and may need a search operation and/or rejection sampling [7; 117]. However, for highly peaked $p(\cdot)$ the increased computational effort is justified by increased acceptance rates [38].

Encorporating more transition types than just births and deaths into the Metropolis–Hastings algorithm 3.1 often leads to better mixing [64; 71; 74; 189; 190; 143]. For instance, suppose a point $x \in \mathcal{X}$ represents a geometric object, so that x contains shape, colour and texture parameters as well as the location of the object. Then a natural transition is to translate the object over a small distance, or to alter one of the object characteristics. Of course, such a change can be accomplished by the death of the original object followed by the birth of the modified one, but, for a plausible configuration, the acceptance probability for the death transition may well be prohibitively small in practice.

Example 3.3 To add a *change* transition to algorithm 3.1, for each $x \in \mathcal{X}$ let $C(x)$ be the set of all $u \in \mathcal{X}$ that can be reached from x by a local change of characteristics. Typically, $C(u)$ will consist of objects that are similar to u. For each configuration $\mathbf{x}$, let $c(\mathbf{x}, x, \cdot) > 0$ be a probability density on $C(x)$ with respect to $\nu(\cdot)$ governing the change of $x \in \mathbf{x}$, and assume the object to be changed is selected randomly. Then detailed balance implies that whenever $p(\mathbf{x})$ and $p((\mathbf{x} \setminus \{x\}) \cup \{u\})$ are strictly positive,

$$\frac{A(\mathbf{x}, (\mathbf{x} \setminus \{x\}) \cup \{u\})}{A((\mathbf{x} \setminus \{x\}) \cup \{u\}, \mathbf{x})} = \frac{p((\mathbf{x} \setminus \{x\}) \cup \{u\})}{p(\mathbf{x})} \frac{c(\mathbf{x}, u, x)}{c(\mathbf{x}, x, u)}.$$

The right hand side is well-defined if we impose the condition that $x \in C(u)$ if and only if $u \in C(x)$.

Example 3.4 Another useful transition type for object processes is to split a large object in two parts [74; 190; 189]. To preserve detailed balance, the possibility of merging two close objects must be included.

As a concrete example, let X be a process of polygonal objects in a compact subset of $\mathbb{R}^2$. Then a split transition begins by selecting two vertices, say v and w, and choosing new vertices v', w' independently, e.g.

from the uniform distribution on a disc centred at v and w respectively. Let v'', w'' be the reflections of v' and w' with respect to v and w. Without loss of generality, we may assume that the edge $v'w'$ lies to the left of $v''w''$, so joining $v'w'$ to the left half of the original polygon and $v''w''$ to the right half completes the proposal.

Writing $s(\mathbf{x}, x, \xi, \eta)$ for the probability density of the proposal to split $x \in \mathbf{x}$ into ξ and η, and $m(\mathbf{y}, \xi, \eta, \zeta)$ for that of merging the objects in $\mathbf{y}$ represented by ξ and η into a single object ζ, we obtain the detailed balance equation

$$p(\mathbf{x}) \, s(\mathbf{x}, x, \xi, \eta) \, A(\mathbf{x}, (\mathbf{x} \setminus \{x\}) \cup \{\xi, \eta\}) =$$

$$p((\mathbf{x} \setminus \{x\}) \cup \{\xi, \eta\}) \, m((\mathbf{x} \setminus \{x\}) \cup \{\xi, \eta\}, \xi, \eta, x) \, A((\mathbf{x} \setminus \{x\}) \cup \{\xi, \eta\}, \mathbf{x})$$

holding for any $x \in \mathbf{x} \in N^{\mathrm{f}}$, $\xi, \eta \in \mathcal{X}$ such that $p(\mathbf{x}), p(\mathbf{x} \setminus \{x\}) \cup \{\xi, \eta\}) > 0$.

To conclude this section, let us look in detail at the hard core process defined by its density

$$p(\mathbf{x}) = \alpha \, \beta^{n(\mathbf{x})} \, \mathbb{I}\{\|x_i - x_j\| > R, i \neq j\}$$

with respect to a unit rate Poisson process on the unit square. Here $n(\mathbf{x})$ as usual denotes the number of points in $\mathbf{x}$, and α, β are strictly positive parameters. Then with probability $1/2$, the Metropolis–Hastings algorithm generates a new point u uniformly on $[0, 1]^2$. If there is a point of $\mathbf{x}$ closer than R from u the proposal is rejected. Otherwise, if $n(\mathbf{x}) \leq \beta - 1$, the proposal is accepted; if $n(\mathbf{x}) \geq \beta$, we accept the new state with probability $\frac{\beta}{n(\mathbf{x})+1}$. In case of a death proposal, each point of $\mathbf{x}$ is selected for removal with equal probability. If $n(\mathbf{x}) \geq \beta$ the proposed death is accepted, while if $n(\mathbf{x}) \leq \beta - 1$ the new state is accepted with probability $\frac{n(\mathbf{x})}{\beta}$.

3.3 Conditional simulation

For many spatial data, the interaction structure is more important than the intensity, and a conditional analysis is performed given the number of points in the data. In contrast to the unconditional case, only transitions leaving the total number of points unchanged are allowed, cf. example 3.3.

As in that example, write $c(\mathbf{x}, x, \cdot) > 0$ for the probability density governing the change of $x \in \mathbf{x}$.

Theorem 3.2 *Let $(\mathcal{X}, d)$ be a complete, separable metric space, $\nu(\cdot)$ a finite, non-atomic Borel measure, and $\pi_\nu(\cdot)$ the distribution of a Poisson process on $\mathcal{X}$ with intensity measure $\nu(\cdot)$. Let $p(\cdot)$ be a probability density with respect to $\pi_\nu(\cdot)$, and condition on the event $\{n(\mathcal{X}) = n\}$. Write $j_n(\cdot, \ldots, \cdot)$ for the probability density governing the locations of the n points, and assume the following reachability conditions hold for all configurations $\mathbf{x}$ consisting of n points, and all $x \in \mathbf{x}$, $u \in \mathcal{X}$:*

- *$c(\mathbf{x}, x, u) > 0 \Leftrightarrow c((\mathbf{x} \setminus \{x\}) \cup \{u\}, u, x) > 0$ whenever $p(\mathbf{x})$ and $p((\mathbf{x} \setminus \{x\}) \cup \{u\})$ are strictly positive;*
- *the Markov process of proposals is irreducible;*
- *the Markov process of proposals is aperiodic or the probability of a self-transition in the Metropolis–Hastings sampler is positive.*

Then the Metropolis–Hastings algorithm 3.1 with acceptance probabilities defined by the Metropolis ratio

$$r(\mathbf{x}, x, u) = \frac{p((\mathbf{x} \setminus \{x\}) \cup \{u\})}{p(\mathbf{x})} \frac{c(\mathbf{x}, u, x)}{c(\mathbf{x}, x, u)}.$$

converges in total variation to the probability distribution specified by $j_n(\cdot, \ldots, \cdot))$ for j_n–almost all initial configurations [64].

Proof. By definition of the Metropolis ratio, $j_n(\cdot, \ldots, \cdot)$ is an invariant density. Let F have positive j_n-mass. Then from any configuration $\mathbf{x} = \{x_1, \ldots, x_n\}$ for which $j_n(x_1, \ldots, x_n) > 0$, the Metropolis–Hastings sampler can reach F with positive probability since the proposal chain is irreducible and the acceptance probability of changing $x_i \in \mathbf{x}$ to u is non-zero whenever $c(\mathbf{x}, x_i, u)$ and $j_n(x_1, \ldots, x_{i-1}, x_{i+1}, \ldots, x_n, u)$ are strictly positive. Furthermore, aperiodicity of the Metropolis–Hastings follows from either the aperiodicity of the proposal process or the self-transitions. Thus, the Metropolis–Hastings algorithm is irreducible and aperiodic on H, the hereditary collection of all configurations $\mathbf{x} = \{x_1, \ldots, x_n\}$ with n points for which $j_n(x_1, \ldots, x_n) > 0$. Hence $j_n(\cdot, \ldots, \cdot)$ defines the unique invariant probability distribution to which the sampler converges in total variation from j_n-almost all initial configurations. $\square$

Example 3.5 Suppose that the proposal density $c(\mathbf{x}, \cdot, \cdot)$ is such that a point x is chosen uniformly from the current configuration $\mathbf{x}$ and replaced by a point u sampled uniformly in some neighbourhood $C(x)$ of x, i.e. [137; 154; 158]

$$c(\mathbf{x}, x, u) = \frac{1}{n} \frac{1}{\nu(C(x))}.$$

The neighbourhoods $C(x)$ must be chosen in such a way that the proposal process is irreducible and aperiodic, for instance $C \equiv \mathcal{X}$.

Another popular choice is the following scheme with alternating births and deaths [174].

Example 3.6 Suppose that the proposal density, as in example 3.5, deletes a point from the current configuration at random but replaces it by sampling from the normalised conditional intensity [174]. The rationale behind this approach is similar to that of example 3.2. Note that

$$c(\mathbf{x}, x, u) = \frac{1}{n} \frac{j_n((\mathbf{x} \setminus \{x\}) \cup \{u\})}{\int_{\mathcal{X}} j_n((\mathbf{x} \setminus \{x\}) \cup \{v\}) \, d\nu(v)}.$$

The Metropolis ratio $r(\mathbf{x}, x, u)$ is identically equal to 1, hence all proposals are accepted.

3.4 Spatial birth-and-death processes

The algorithms considered in examples 3.2 and 3.6 can be seen as discretisations of a *spatial birth-and-death process* [166]. Originally phrased in continuous time, running a spatial birth-and-death process is perhaps the oldest technique for simulating a Markov point process advocated in the statistical literature [7; 11; 39; 46; 116; 141; 174; 176; 179]. All transition proposals are accepted with probability 1, but the process stays in state $X^{(n)}$ for an exponentially distributed random sojourn time $T^{(n)}$. As for the Metropolis–Hastings procedure of the previous section, the transitions from a state $\mathbf{x}$ are births and deaths, with respective rates $b(\mathbf{x}, u)$ $(u \in \mathcal{X})$

and $d(\mathbf{x}, x)$, $x \in \mathbf{x}$. The total transition rate is the sum of the total birth rate

$$B(\mathbf{x}) = \int_{\mathcal{X}} b(\mathbf{x}, u) \, d\nu(u)$$

and the total death rate

$$D(\mathbf{x}) = \sum_{x \in \mathbf{x}} d(\mathbf{x}, x);$$

the sojourn time $T^{(n)}$ in state $X^{(n)} = \mathbf{x}$ has mean $1/(B(\mathbf{x}) + D(\mathbf{x}))$.

Summarising, to simulate the spatial birth-and-death process we generate the successive states $X^{(n)}$ and the sojourn times $T^{(n)}$ as follows.

Algorithm 3.2   Initialise $X^{(0)} = \mathbf{x}_0$ for some configuration $\mathbf{x}_0 \in N^f$ with $p(\mathbf{x}_0) > 0$. For $n = 0, 1, \ldots$, if $X^{(n)} = \mathbf{x}$

- $T^{(n)}$ is exponentially distributed with mean $1/(D(\mathbf{x}) + B(\mathbf{x}))$ independent of other sojourn times and of past states;
- the next transition is a *death* with probability $D(\mathbf{x})/(D(\mathbf{x}) + B(\mathbf{x}))$, obtained by deleting one of the current points x with probability

$$\frac{d(\mathbf{x}, x)}{D(\mathbf{x})};$$

- with probability $B(\mathbf{x})/(D(\mathbf{x}) + B(\mathbf{x}))$ the next transition is a *birth*, generated by choosing u according to the probability density

$$\frac{b(\mathbf{x}, u)}{B(\mathbf{x})}$$

  with respect to the reference measure $\nu(\cdot)$ and adding u to the state $\mathbf{x}$.

The detailed balance equations are given by

$$b(\mathbf{x}, u) \, p(\mathbf{x}) = d(\mathbf{x} \cup \{u\}, u) \, p(\mathbf{x} \cup \{u\}) \tag{3.5}$$

for any $\mathbf{x} \in N^f$, $u \in \mathcal{X}$ such that $p(\mathbf{x} \cup \{u\}) > 0$. Thus, once the death rates are specified, the birth rates can be computed immediately.

Example 3.7 A common choice [174] is to take a constant death rate $d \equiv 1$ yielding

$$b(\mathbf{x}, u) = \frac{p(\mathbf{x} \cup \{u\})}{p(\mathbf{x})},$$

the Papangelou conditional intensity at u given $\mathbf{x}$. Note that once the process is in the hereditary set $\{\mathbf{x} \in N^f : p(\mathbf{x}) > 0\}$ of feasible configurations, it almost surely never leaves it.

Since the process evolves in continuous time, conditions have to be imposed to avoid 'explosion', i.e. an infinite number of transitions in finite time [166].

Theorem 3.3 *Let $(\mathcal{X}, d)$ be a complete, separable metric space, $\nu(\cdot)$ a finite, non-atomic Borel measure, and $\pi_\nu(\cdot)$ the distribution of a Poisson process on $\mathcal{X}$ with intensity measure $\nu(\cdot)$. Let $p(\cdot)$ be a probability density with respect to $\pi_\nu(\cdot)$. For each $n = 0, 1, \ldots$ define $\beta_n = \sup_{n(\mathbf{x})=n} B(\mathbf{x})$ and $\delta_n = \inf_{n(\mathbf{x})=n} D(\mathbf{x})$ and assume that $\delta_n > 0$ for all $n \geq 1$. Assume either $\beta_n = 0$ for all sufficiently large $n \geq 0$, or $\beta_n > 0$ for all $n \geq 1$, and that both the following hold:*

$$\sum_{n=2}^{\infty} \frac{\beta_1 \cdots \beta_{n-1}}{\delta_1 \cdots \delta_n} < \infty$$

$$\sum_{n=1}^{\infty} \frac{\delta_1 \cdots \delta_n}{\beta_1 \cdots \beta_n} = \infty.$$

Then there exists a unique spatial birth-and-death process for which $b(\cdot, \cdot)$ and $d(\cdot, \cdot)$ are the transition rates; this process has a unique equilibrium distribution to which it converges in distribution from any initial state.

Proof. (Sketch)
Consider the simple birth-and-death process with state space $\{0, 1, 2, \ldots\}$, birth rates β_n, and death rates δ_n, $n \in \mathbb{N}_0$. In order to find an invariant distribution $(\pi_n)_{n \geq 0}$, consider the balance equations

$$\beta_0 \pi_0 = \delta_1 \pi_1$$
$$(\beta_n + \delta_n) \pi_n = \beta_{n-1} \pi_{n-1} + \delta_{n+1} \pi_{n+1} \qquad n \geq 1.$$

Solving these equations recursively, one obtains $\delta_n \pi_n = \beta_{n-1} \pi_{n-1}$ for $n \geq 1$, and hence

$$\pi_n = \frac{\beta_{n-1}}{\delta_n} \pi_{n-1} = \cdots = \frac{\beta_{n-1} \cdots \beta_0}{\delta_n \cdots \delta_1} \pi_0.$$

In order for $(\pi_n)_n$ to be well-defined, we have to require $\delta_n > 0$ for all $n \geq 1$. If $\beta_n = 0$ for all sufficiently large n, so is π_n and hence there is a unique invariant probability mass function. Otherwise,

$$1 = \pi_0 \left[1 + \frac{\beta_0}{\delta_1} + \beta_0 \sum_{n=2}^{\infty} \frac{\beta_1 \cdots \beta_{n-1}}{\delta_1 \cdots \delta_n} \right]$$

and the infinite series in the last term must be finite.

To guarantee the existence of a Markov process with the given birth and death rates and avoid explosion, the Kolmogorov equations [37] must have a unique solution. A sufficient condition is that either $\beta_n = 0$ for all sufficiently large n, or that $\beta_n > 0$ for all $n \geq 1$ and

$$\sum_{n=1}^{\infty} \frac{\delta_1 \cdots \delta_n}{\beta_1 \cdots \beta_n} = \infty.$$

Finally it can be shown [166] that the spatial birth-and-death process can be coupled to the simple birth-and-death process in such a way that existence and limit results carry over.   □

Example 3.8 Let $p(\cdot)$ be a Markov point process density and assume its conditional intensity is bounded, that is $\lambda(u; \mathbf{x}) \leq \beta$ for some constant $\beta > 0$. Such densities are called *locally stable*.   Then, for the constant death rate sampler [174] introduced in example 3.7,

$$B(\mathbf{x}) = \int_{\mathcal{X}} \lambda(u; \mathbf{x}) \, d\nu(u) \leq \beta \nu(\mathcal{X})$$

and $\beta_n \leq \beta \nu(\mathcal{X})$ for all n. Moreover $D(\mathbf{x}) = n(\mathbf{x})$, implying that $\delta_n = n$. Therefore, if $\beta_n > 0$ for all $n \geq 1$

$$\sum_{n=2}^{\infty} \frac{\beta_1 \cdots \beta_{n-1}}{\delta_1 \cdots \delta_n} \leq \sum_{n=2}^{\infty} \frac{(\beta \nu(\mathcal{X}))^{n-1}}{n!} < \infty$$

and

$$\sum_{n=1}^{\infty} \frac{\delta_1 \cdots \delta_n}{\beta_1 \cdots \beta_n} \geq \sum_{n=1}^{\infty} \frac{n!}{(\beta \, \nu(\mathcal{X}))^n} = \infty.$$

By theorem 3.3 there exists a unique spatial birth-and-death process for which $b(\mathbf{x}, u) = \lambda(u; \mathbf{x})$ and $d \equiv 1$ are the transition rates, and this process converges to $p(\cdot)$ from any initial state $\mathbf{x}$ with $p(\mathbf{x}) > 0$.

Generally speaking, the total birth rate $B(\mathbf{x})$ is difficult to compute. For locally stable target distributions however, thinning can be used to implement algorithm 3.2. Indeed, since the total birth rate is bounded by $\beta \nu(\mathcal{X})$ for some $\beta > 0$, computation of $B(\mathbf{x})$ can be avoided by running a spatial birth-and-death process with birth rate β, and accepting any birth transition from $\mathbf{x}$ to $\mathbf{x} \cup \{u\}$ with probability $\lambda(u; \mathbf{x})/\beta$. This strategy works best if the interaction is mild. If the conditional intensity is highly peaked, the acceptance probabilities may well be too small to be workable and it is a good idea to include a search step or more sophisticated rejection sampling techniques [7; 117].

3.5 Exact simulation

The MCMC algorithms presented in sections 3.2–3.4 in principle must be run for an infinite length of time in order to reach equilibrium. In practice, of course, the algorithm is run for a finite time, and convergence is monitored by plotting time series of diagnostic statistics. Such plots can be misleading. For instance for a multi-modal target distribution, an MCMC sampler may have difficulty jumping between modes, which results in a seemingly stable behaviour that in fact only represents the distribution around a single mode.

Recently Propp and Wilson [167] showed that sometimes it is possible to let the Markov process itself indicate when its equilibrium is reached. To explain their main ideas, suppose for the moment that the state space is a finite set $S = \{s_1, \ldots, s_m\}$ and the goal is to sample from a positive probability mass function $\pi(\cdot)$ on S by running a Markov chain $(X_n)_n$ into equilibrium. This Markov chain may be represented by a *transition function*

$\phi(\cdot, \cdot)$, so that

$$X_{n+1} = \phi(X_n, V_n) \tag{3.6}$$

for some random vectors V_n. Usually in MCMC, the transition function is implicit, but it plays a major role in exact simulation algorithms.

Imagine m parallel chains $(X_n(s_i))_n$ are run from some time $-T < 0$ until 0, one for each initial state $s_i \in S$. The chains are coupled so that each of them uses *the same* V_ns. Hence, if at any stage the initial state influence has worn off, i.e. $X_n \equiv x$ (say) for some $n \in \{-T, \dots, 0\}$, then all sample paths remain identical until time 0. We say that the m chains have coalesced. Consequently, had we sampled for an infinite amount of time, the same result X_0 would have been obtained, and X_0 is an exact sample from the equilibrium distribution $\pi(\cdot)$.

Of course, it is not practically feasible to run many parallel chains. However, if the state space is partially ordered, with a maximum $s_{\max}$ and minimum $s_{\min}$, and moreover the transition function respects the order in the sense that $s \leq s'$ implies $\phi(s, \cdot) \leq \phi(s', \cdot)$, then there is no need to consider all initial states. Indeed the two chains starting in $s_{\max}$ and $s_{\min}$ suffice, as all other sample paths are sandwiched between the upper and loser chains [167].

Algorithm 3.3   Initialise $T = 1$. Let $\phi(\cdot, \cdot)$ be a transition function on the partially ordered set $S = \{s_1, \dots, s_m\}$ with maximum $s_{\max}$ and minimum $s_{\min}$, and V_n, $n = -1, -2, \dots$ independently, identically distributed random vectors as in (3.6). Repeat

- initialise $U_{-T}(-T) = s_{\max}$, $L_{-T}(-T) = s_{\min}$;
- for $n = -T$ to -1, set $U_{-T}(n + 1) = \phi(U_{-T}(n), V_n)$ and $L_{-T}(n + 1) = \phi(L_{-T}(n), V_n)$;
- if $U_{-T}(0) = L_{-T}(0)$, return the common value; otherwise set $T := 2T$;

until the chains have coalesced.

The upper and lower chains are coupled from the past: when extending backwards from time $-T$ to $-2T$, the values of V_n, $n = -T, \dots, -1$ are reused in both chains. It is natural to ask whether coupling into the future would work as well. Unfortunately, this is not the case [167].

Example 3.9 Let S be the set $\{1, 2\}$ and define a Markov chain [103] on S as follows. From state 1, the chain moves to either state with equl probability $1/2$, while from state 2 it always jumps to 1. Then, coalesce will occur for the first time in state 1, but the invariant measure assigns mass $1/3$ to state 2.

The following theorem [167] gives sufficient conditions for the output of algorithm 3.3 to be an unbiased sample from the target distribution.

Theorem 3.4 *Let $\pi(\cdot)$ be a strictly positive probability mass function on a finite partially ordered set $(S, \leq)$ with minimum $s_{\min}$ and maximum $s_{\max}$. Let $\phi(\cdot, \cdot)$ be the transition function of an irreducible, aperiodic Markov chain with invariant distribution $\pi(\cdot)$ that respects the partial order. Then whenever the Propp-Wilson algorithm 3.3 terminates almost surely, it outputs an unbiased sample from $\pi(\cdot)$.*

Proof. Let $\pi'(\cdot)$ be the distribution of the output of Algorithm 3.3 and write I for the stopping time denoting that value of T for which coalescence first occurs. For fixed $\epsilon > 0$, since by assumption $I < \infty$ a.s. we can pick t such that $\mathbf{P}(I > t) \leq \epsilon$. Define a third Markov chain $X_{-t}(n)$ from time $-t$ onwards in the same way as $U_{-t}(n)$, $L_{-t}(n)$ except that $X_{-t}(-t)$ is initialised by a sample from $\pi(\cdot)$, independently of all other random vectors. Since $\pi(\cdot)$ is the invariant distribution, $X_{-t}(0)$ too is distributed according to $\pi(\cdot)$. Furthermore, $U_{-t}(0) = L_{-t}(0) = X_{-t}(0)$ on the event $\{I \leq t\}$, hence the total variation distance between $\pi(\cdot)$ and $\pi'(\cdot)$ satisfies

$$\|\pi - \pi'\| \leq \mathbf{P}(I > t) \leq \epsilon.$$

Since ϵ was chosen arbitrarily, $\|\pi - \pi'\| = 0$, hence $\pi(\cdot) = \pi'(\cdot)$. $\square$

Not all choices of the transition function result in an almost surely coalescing chain.

Example 3.10 Further to example 3.9, let $S = \{1, 2\}$ be ordered by $1 < 2$ and equipped with a probability measure $\pi(\cdot)$ assigning equal probability

$1/2$ to each member. Then the transition matrix defined by $p_{ij} = \frac{1}{2}$ ($i = 1, 2$) is irreducible, aperiodic and has invariant measure $\pi(\cdot)$. This chain can be implemented by the transition function

$$\phi(X_n, (V_{n1}, V_{n2})) = \begin{cases} 1 & \text{if } V_{n,X_n} > \frac{1}{2} \\ 2 & \text{if } V_{n,X_n} \leq \frac{1}{2} \end{cases}$$

where the marginal distribution of V_{ni} ($i = 1, 2$) is that of a uniform distribution on $[0, 1]$. The joint distribution of the random vectors (V_{n1}, V_{n2}) can be chosen arbitrarily. Thus, let V_{11} be a uniformly distributed random variabe. If we set $V_{12} = 1 - V_{11}$, the chain never coalesces. The more natural choice $V_{12} = V_{11}$ on the other hand leads to a Markov chain that almost surely merges in a single step. If the components V_{n1}, V_{n2} are independent, algorithm 3.3 terminates almost surely, but in general – as here – this is not the most efficient choice [167].

Algorithm 3.3 can be adapted to transition functions that reverse the order by simply using the current value of the other chain when updating: $U_{-T}(n+1) = \phi(L_{-T}(n), V_n)$ and $L_{-T}(n+1) = \phi(U_{-T}(n), V_n)$. The proof of theorem 3.4 carries over with straightforward modifications [78].

How do the above ideas generalise to point processes? Of course there is no problem in conceptualising a family of coupled processes, but some care is required to ensure that infinitely many such chains eventually coalesce [148].

Spatial birth-and-death processes (cf. section 3.4) are particularly a-menable to coupling. As before, let $p(\cdot)$ be a point process density on a complete, separable metric space $\mathcal{X}$. The set N^f of finite subsets of $\mathcal{X}$ is partially ordered by inclusion. The empty set is the minimal element, but there is no maximum, so that a stochastic maximum will be used in algorithm 3.4 below. Furthermore, assume that the conditional intensity can be bounded uniformly in $\mathbf{x} \in N^f$, that is

$$0 \leq \lambda(u; \mathbf{x}) \leq \lambda(u)$$

for some integrable function $\lambda(\cdot)$.

Since birth-and-death processes are defined in continuous time, the role of the transition function in algorithm 3.3 is taken over by the transition rates. For the constant death rate sampler introduced in example 3.7, since the birth rate is proportional to the conditional intensity, the process respects the inclusion order if

$$\lambda(u; \mathbf{x}) \leq \lambda(u; \mathbf{y}) \qquad (3.7)$$

whenever $\mathbf{x} \subseteq \mathbf{y}$. Point processes satisfying (3.7) are called *attractive*. As in the discussion at the end of section 3.4, the sampler may be implemented by thinning a spatial birth-and-death process with birth rate $\lambda(\cdot)$ and death rate 1.

The following algorithm [9], generalising an exact sampler for locally stable point processes [103; 105], implements these ideas.

Algorithm 3.4 Let X be a Markov point process on $\mathcal{X}$, and assume its Papangelou conditional intensity is bounded by the integrable function $\lambda(\cdot)$. Let $V_{t,u}$, $t \leq 0$, $u \in \mathcal{X}$, be a family of independent, uniformly distributed random variables on $(0,1)$. Initialise $T = 1$, and let $D(0)$ be a sample from a Poisson process with intensity function $\lambda(\cdot)$. Repeat

- extend $D(\cdot)$ backwards until time $-T$ by means of a spatial birth-and-death process with birth rate $\lambda(\cdot)$ and death rate 1;
- generate a lower process $L_{-T}(\cdot)$ and an upper process $U_{-T}(\cdot)$ on $[-T, 0]$ as follows:

    - initialise $L_{-T}(-T) = \emptyset$, $U_{-T}(-T) = D(-T)$;
    - to each forward transition time $t \in (-T, 0]$ of $D(\cdot)$ correspond updates of the upper and lower processes;
    - in case of a death (i.e. a backwards birth), say $D(t) = D(t-) \setminus \{d\}$ where $D(t-)$ denotes the state just prior to time t, the point d is deleted from $L_{-T}(t-)$ and $U_{-T}(t-)$ as well;
    - in case of a birth, say $D(t) = D(t-) \cup \{u\}$, the point u is added to $U_{-T}(t-)$ only if $V_{t,u} \leq \lambda(u; U_{-T}(t-))/\lambda(u)$; similarly, u is added to $L_{-T}(t-)$ only if $V_{t,u} \leq \lambda(u; L_{-T}(t-))/\lambda(u)$;

- if $U_{-T}(0) = L_{-T}(0)$, return the common value $U_{-T}(0)$; otherwise set $T := 2T$;

until the upper and lower processes have coalesced.

As for the discrete Propp-Wilson algorithm 3.3, for *repulsive* point processes whose conditional intensity satisfies

$$\lambda(u; \mathbf{x}) \leq \lambda(u; \mathbf{y})$$

whenever $\mathbf{y} \subseteq \mathbf{x}$, the algorithm can easily be modified [103]. For instance, the proposal to add a point u to $U_{-T}(t-)$ is accepted if $V_{t,u} \leq \lambda(u; L_{-T}(t-))/\lambda(u)$.

Theorem 3.5 *Let $(\mathcal{X}, d)$ be a complete, separable metric space, $\nu(\cdot)$ a finite, non-atomic Borel measure, and $\pi_\nu(\cdot)$ the distribution of a Poisson process on $\mathcal{X}$ with intensity measure $\nu(\cdot)$. Let $p(\cdot)$ be an attractive probability density with respect to $\pi_\nu(\cdot)$ such that for all $\mathbf{x} \in N^f$, $u \notin \mathbf{x}$,*

$$\lambda(u; \mathbf{x}) = \frac{p(\mathbf{x} \cup \{u\})}{p(\mathbf{x})} \leq \lambda(u)$$

for some Borel measurable, integrable function $\lambda : \mathcal{X} \to \mathbb{R}^+$. Then the coupling from the past algorithm 3.4 almost surely terminates and outputs an unbiased sample from $p(\cdot)$.

Proof. First, note that the dominating process $D(\cdot)$ is in equilibrium, its distribution being that of a Poisson process with intensity function $\lambda(\cdot)$. Clearly, for all $T > 0$,

$$\emptyset = L_{-T}(-T) \subseteq U_{-T}(-T) = D(-T).$$

Since the process is attractive, the updates respect the inclusion order. Hence $L_{-T}(t) \subseteq U_{-T}(t)$ for all $t \in [-T, 0]$. Moreover, the processes funnel, i.e.

$$L_{-T}(t) \subseteq L_{-S}(t) \subseteq U_{-S}(t) \subseteq U_{-T}(t) \qquad (3.8)$$

whenever $-S \leq -T \leq t \leq 0$. The first inclusion can be verified by noting that $L_{-T}(-T) = \emptyset \subseteq L_{-S}(-T)$ and recalling that the transitions respect the inclusion order. Since $U_{-T}(-T) = D(-T) \supseteq U_{-S}(-T)$, the last inclusion in (3.8) follows. If $L_{-T}(t_0) = U_{-T}(t_0)$ for some $t_0 \in [-T, 0]$, as the processes are coupled, $L_{-T}(t) = U_{-T}(t)$ for all $t \in [t_0, 0]$.

Next, set $X_{-T}(-T) = \emptyset$ and define a process $X_{-T}(\cdot)$ on $[-T, 0]$ in analogy to the upper and lower processes, except that a birth is accepted if $V_{t,u} \leq \lambda(u; X(t-))/\lambda(u)$. In other words, $X_{-T}(\cdot)$ exhibits the dynamics of a spatial birth-and-death process with birth rate $\lambda(\cdot; \cdot)$, the conditional intensity, and with death rate 1. By assumption, $\lambda(u; \mathbf{x}) \leq \lambda(u)$, hence the total birth rate $B(\mathbf{x})$ from state $\mathbf{x}$ is bounded by $\Lambda = \int_{\mathcal{X}} \lambda(u)\, d\nu(u)$, and theorem 3.3 implies that the process converges in distribution to its equilibrium distribution defined by $p(\cdot)$. Returning to $X_{-T}(\cdot)$, note that the monotonicity assumption implies $L_{-T}(0) \subseteq X_{-T}(0) \subseteq U_{-T}(0)$, so that – provided the sampler terminates almost surely – with probability 1 the limit $\lim_{T\to\infty} X_{-T}(0)$ is well-defined. Extend $D(\cdot)$ forwards in time by means of a spatial birth-and-death process with birth rate $\lambda(\cdot)$ and death rate 1. Since $D(\cdot)$ is in equilibrium, $X_{-T}(0)$ has the same distribution as its forward extension (coupled to the dominating process as before) over a time period of length T. We conclude that $L_{-T}(0)$ is an unbiased sample from $p(\cdot)$.

It remains to show that coalescence occurs almost surely. Note that $\pi_\lambda(\emptyset) = \exp\left[-\int_{\mathcal{X}} \lambda(u)\, d\nu(u)\right] > 0$, where $\pi_\lambda(\cdot)$ is the distribution of a Poisson process with intensity function $\lambda(\cdot)$. Set, for $n \in \mathbb{N}_0$, $E_n = \mathbb{I}\{D(-n) \neq \emptyset\}$. Now $(E_n)_n$ is an irreducible aperiodic Markov chain on $\{0, 1\}$ for which the equilibrium probability $\pi_\lambda(\emptyset)$ of state 0 is strictly positive. Hence state 0 will be reached with probability 1, which implies that the dominating process $D(t)_{t\leq 0}$ will almost surely be empty for some t. But then (3.8) and the coupling imply that the algorithm terminates almost surely, and the proof is complete. $\qquad\square$

Example 3.11 Consider the hard core process with density (2.1). Write $\mathcal{Y}_\mathbf{x} = \{u \in \mathcal{X} : \|x - u\| > R \text{ for all } x \in \mathbf{x}\}$. Then the conditional intensity can be written as

$$\lambda(u; \mathbf{x}) = \beta\, \mathbb{I}\{u \in \mathcal{Y}_\mathbf{x}\}$$

which is bounded by β. Furthermore, since for $\mathbf{y} \subseteq \mathbf{x}$ clearly $\mathcal{Y}_\mathbf{y} \supseteq \mathcal{Y}_\mathbf{x}$, the hard core process is repulsive. Therefore algorithm 3.3 can be used to obtain samples from $p(\cdot)$. The stochastic upper bound $D(\cdot)$ is a Poisson process with intensity β. All forward death transitions in $D(\cdot)$ are mimicked, whilst a point u born into $D(t-)$ at time t is added to the upper process if and only if u does not violate the hard core distance R in $L(t-)$; the L-process is updated similarly. Note that the points in the upper process not necessarily keep a distance R from each other, but that the points in the lower process do. When the two processes meet, a sample from the hard core model is obtained.

Algorithm 3.3 was used to obtain the realisation with $\beta = 100$ and $R = 0.05$ displayed in Figure 2.1. Coalescence took place for $T = -16$.

3.6 Auxiliary variables and the Gibbs sampler

The goal of introducing auxiliary variables is to counteract strong interactions in the original model formulation that result in slowly mixing Markov chains [22]. Sometimes, the new variables can be given a physical interpretation, more often they are quite abstract. Always, the result should be a chain that allows for more substantial state changes than the standard Metropolis or spatial birth-and-death sampler.

Let X be a point process on a complete, separable metric space $\mathcal{X}$, specified by its density $p(\cdot)$ with respect to a Poisson process with finite, non-atomic intensity measure $\nu(\cdot)$. We will write Y for the auxiliary variables, $\mathcal{Y}$ for its state space. To obtain a joint probabilistic model for X and Y, we need to specify the conditional distribution of Y given X. For simplicity, we shall assume that this conditional distribution has a density $p(\mathbf{y} \mid \mathbf{x})$ with respect to a probability law $\pi(\cdot)$ on $\mathcal{Y}$.

An auxiliary variables technique alternatingly updates the two types of variables, the original and the auxiliary ones. In order to converge to the desired limit distribution given by $p(\mathbf{x})$, each cycle must preserve $p(\cdot)$. For example, a *Gibbs approach* would be to use the conditional distributions as in the following algorithm.

Algorithm 3.5 Initialise $X_0 = \mathbf{x}_0$ for some configuration $\mathbf{x}_0 \in N^f$ with $p(\mathbf{x}_0) > 0$. For $n = 0, 1, \ldots$, if $X_n = \mathbf{x}$

- draw a new value $\mathbf{y}$ of the auxiliary variable according to $p(\mathbf{y} \mid \mathbf{x})$;
- sample a new X-state from $p(\cdot \mid \mathbf{y})$.

Each cycle of algorithm 3.5 preserves the target distribution hence, provided the usual irreducibility and aperiodicity conditions hold, algorithm 3.5 converges in total variation to $p(\cdot)$ for almost all initial states. If some partial order structure can be demonstrated for the Gibbs updates, coupling from the past ideas apply (cf. section 4.2–4.3).

General rules for choosing the auxiliary variables are hard to give, depending as they do on the structure of the model in hand. The following example illustrates a fairly general construction for Markov point processes with interactions of bounded order, that is, whose interaction function in the Hammersley–Clifford factorisation is identically 1 for cliques with cardinality higher than some given $K \geq 2$.

Example 3.12 Consider a Markov point process X with density

$$p(\mathbf{x}) = \alpha \, \beta^{n(\mathbf{x})} \prod_{k=2}^{K} \Phi_k(\mathbf{x})$$

where $\beta > 0$ is an intensity parameter, and $n(\mathbf{x})$ denotes the number of points in $\mathbf{x}$. Furthermore, for $k = 2, \ldots, K$,

$$\Phi_k(\mathbf{x}) = \prod_{\mathbf{z} \subseteq \mathbf{x} : n(\mathbf{z}) = k} \phi(\mathbf{z})$$

is the product of the interaction function over cliques consisting of exactly k points.

For each order $k = 2, \ldots, K$, conditionally on $X = \mathbf{x}$, let Y_k be uniformly distributed on the interval $[0, \Phi_k(\mathbf{x})]$. Thus, the joint distribution of X and the auxiliary $(K-1)$-vector $Y = (Y_2, \ldots, Y_{K-1})$ can be described by its density [22; 52]

$$p(\mathbf{x}, \mathbf{y}) = \alpha' \, \beta^{n(\mathbf{x})} \prod_{k=1}^{K} \mathbb{I}\{y_k \leq \Phi_k(\mathbf{x})\}$$

with respect to the product of a unit rate Poisson process and $(K-1)$-fold Lebesgue measure. Hence the conditional distribution of X given $Y = \mathbf{y}$ is that of a Poisson process of rate β under the restriction that $\Phi_k(X) > y_k$, $k = 2, \ldots, K$.

The most satisfying examples of auxiliary variable techniques are those for which Y reflects a salient model feature.

Example 3.13 Consider the area-interaction process (example 2.11) with density

$$p(\mathbf{x}) = \alpha \beta_1^{n(\mathbf{x})} \exp[-\beta_2 \nu(U_{\mathbf{x}})] \tag{3.9}$$

where $U_{\mathbf{x}} = \bigcup_{x \in \mathbf{x}} B(x, R)$ denotes the union of balls around points of $\mathbf{x}$, $n(\mathbf{x})$ the cardinality of $\mathbf{x}$, and $\beta_1, \beta_2 > 0$ are model parameters. As demonstrated in section 2.3, (3.9) is the marginal distribution of the first component in a bivariate model with density

$$p(\mathbf{x}, \mathbf{y}) = \alpha \beta_1^{n(\mathbf{x})} \beta_2^{n(\mathbf{y})} \, \mathbb{I}\{d(\mathbf{x}, \mathbf{y}) > R\} \tag{3.10}$$

with respect to the product of two unit intensity Poisson processes. Although the area-interaction model is locally stable, and may be simulated by any of the general techniques discussed in the previous sections, the relationship to the mixture model can be used to devise a Gibbs sampler with the second component pattern as the auxiliary variable [77; 32].

If the current state is $(\mathbf{x}, \mathbf{y})$, a cycles of the two-component Gibbs sampler is as follows:

- sample $\mathbf{y}$ according to a Poisson process of intensity β_2 on $\mathcal{X} \setminus U_{\mathbf{x}}$;
- sample a new X-configuration $\mathbf{x}'$ according to a Poisson process of intensity β_1 on $\mathcal{X} \setminus U_{\mathbf{y}}$.

The Poisson processes involved are easily implemented by generating a Poisson process on $\mathcal{X}$ and deleting the points that fall in $U_{\mathbf{x}}$ respectively $U_{\mathbf{y}}$.

The introduction of the Y-component succeeds in altering the whole pattern $\mathbf{x}$ in a single cycle, rather than updating a point at a time as in the Metropolis–Hastings or birth-and-death techniques of sections 3.2–3.4. The realisation shown in Figure 2.2 was obtained using the Gibbs sampler (modified to an exact sampler using Propp-Wilson style ideas, cf. chapter 4).

The Gibbs sampler is not the only valid choice in working with auxiliary variables; indeed any strategy that preserves the required equilibrium distribution may be considered [22]. For instance, in a Swendsen-Wang style method [211], all points in a 'connected component' are updated simultaneously. Sometimes these components are natural model ingredients as in example 3.13, more generally bond variables (example 3.12) can be used to connect the points.

Example 3.14 Let

$$p(\mathbf{x}) = \alpha \, \beta^{n(\mathbf{x})} \prod_{u,v \in \mathbf{x}: u \sim v} \phi(\|u - v\|)$$

be the density of a pairwise interaction process on a rectangle $\mathcal{X} \subset \mathbb{R}^d$ wrapped on a torus, and suppose we are interested in a sample consisting of exactly $n \in \mathbb{N}_0$ points. Provided the interaction function $\phi(\cdot)$ can be bounded away from 0, that is $\phi_* = \inf_{\mathbf{x} \in N^f} \phi(\mathbf{x}) > 0$, bond variables between $u, v \in \mathbf{x}$ can be defined as independent alternatives B_{uv},

$$\mathbf{P}(B_{uv} = 0 \mid \mathbf{x}) = \frac{\phi_*}{\phi(|u - v|)} = 1 - \mathbf{P}(B_{uv} = 1 \mid \mathbf{x}).$$

Note that the stronger the inhibition between u and v, the smaller the probability of a bond $B_{uv} = 1$ between them.

The Swendsen–Wang–Cai algorithm [28] updates componentwise as follows. First, identify the connected components of the current configuration $\mathbf{x}$ formed by linking those points whose bond is 'on', i.e. those $u, v \in \mathbf{x}$ for which $B_{uv} = 1$. Then translate each component independently and randomly over $\mathcal{X}$. Clearly this procedure is irreducible and aperiodic; since the algorithm updates all points simultaneously, it leads to better mixing than algorithms based on births and deaths only.

Example 3.15 Following example 3.13, consider the area-interaction point process with density (3.9) for $\beta_1 = \beta_2 = \beta$ and its associated bivariate mixture model (3.10). In order to define a Swendsen–Wang algorithm, recall from section 2.3 that the locations $\mathbf{z}$ of all points regardless of the component label constitute a continuum random cluster process, and given $\mathbf{z}$, the labels are assigned independently to each component of $\bigcup_{z \in \mathbf{z}} B(z, R/2)$ with probability $1/2$ for each label. Thus, if the current configuration of the first component is $\mathbf{x}$, a Swendsen–Wang update [32; 77] consists of

- sample $\mathbf{y}$ from a Poisson process on $\mathcal{X} \setminus U_{\mathbf{x}}$ of intensity β;

- sample a new partition $(\mathbf{x}', \mathbf{y}')$ by assigning a random mark to each component of $\bigcup z \in \mathbf{x} \cup \mathbf{y} B(z, R/2)$.

For large values of the intensity parameter β, although (3.10) is symmetric in the two components, realisations tend to be dominated by a single component [31]. The Swendsen–Wang algorithm, since it can change component labels, more easily jumps out of such realisations than the Gibbs sampler of example 3.13, leading to improved mixing.

To conclude our discussion of Monte Carlo techniques, consider sampling from a mixture model

$$p(\mathbf{x}) = \sum_{i=1}^{m} w_i p_i(\mathbf{x}) \tag{3.11}$$

where the point process densities $p_i(\cdot)$ are defined on the same probability space and the weights $w_i > 0$ satisfy $\sum_{i=1}^{m} w_i = 1$. Usually, it is not so much the mixture distribution itself that is of interest. More often, (3.11) is used when simultaneous sampling from several probability distributions is needed, for instance for a range of parameter values in some exponential family, or when $p_1(\cdot)$ is of prime importance but direct sampling from it is hard, e.g. due to very strong interactions or severe multimodality. In cases like this, the other mixture components in effect are auxiliary variables, introduced to improve mixing and avoid getting stuck in a local optimum. The success of the method relies on the choice of the weights w_i and the auxiliary distributions $p_i(\cdot)$, a full treatment of which lies beyond the scope of this book. The interested reader is referred to e.g. [61; 76; 128; 144] for more details.

3.7   Maximum likelihood estimation

In this section we will review estimation techniques for Markov point processes. In the early days of spatial statistics, inference was mostly of a non-parametric nature and based on summary statistics such as Ripley's K-function [46; 174; 176; 179]. Later on, following the introduction of parametric models, Takacs-Fiksel [53; 54; 212; 213], stochastic approximation [13; 19; 147; 154; 155; 158; 162] and pseudo-likelihood [20; 24; 97; 95; 69; 196] techniques were developed. The huge progress in Markov chain Monte Carlo techniques has facilitated maximum likelihood estimation [61; 65; 63;

64]. In this section, we will trace our steps back in time and describe the maximum likelihood approach first. Subsequent sections are devoted to alternatives based on the Nguyen–Zessin formula (1.16) and various summary statistics.

The general set up will be a point process specified by its density $p(\cdot; \theta)$ with respect to the distribution $\pi(\cdot)$ of a Poisson process with finite non-atomic intensity measure $\nu(\cdot)$, $0 < \nu(\mathcal{X}) < \infty$. The density is given in a parametric form, with the parameter θ lying in some space Θ.

Example 3.16 For a homogeneous Poisson process of unknown intensity θ,

$$\log p(\mathbf{x}; \theta) = (1 - \theta)\,\nu(\mathcal{X}) + n(\mathbf{x}) \log \theta$$

where $n(\mathbf{x})$ is the cardinality of $\mathbf{x}$. Differentiating with respect to θ and equating the result to zero yields

$$-\nu(\mathcal{X}) + \frac{n(\mathbf{x})}{\theta} = 0$$

hence an estimator for the intensity is $\hat{\theta} = \frac{n(\mathbf{x})}{\mu(\mathcal{X})}$, the observed number of points per unit area. To verify that $\hat{\theta}$ is indeed the maximum likelihood estimator, note that the second order derivative of $p(\mathbf{x}; \theta)$ is strictly positive provided $n(\mathbf{x}) \neq 0$. Hence $\hat{\theta}$ is the unique solution to the maximum likelihood equation.

Note that for an inhomogeneous Poisson process with density

$$\log p(\mathbf{x}; \lambda) = \int_{\mathcal{X}} (1 - \lambda(x))\,d\nu(x) + \sum_{x \in \mathbf{x}} \log \lambda(x)$$

the above approach is still valid if the intensity function $\lambda(\cdot) = \lambda(\cdot; \theta)$ is given in a parameterised form. More generally, the parameter $\lambda(\cdot)$ is infinite dimensional and non-parametric techniques are called for. For instance [89], $\lambda(\cdot)$ may be approximated by a function that is constant on the members of some partition of $\mathcal{X}$, or by a kernel estimator [47; 50; 199].

Example 3.17 Define a hard core process by restricting a homogeneous Poisson process to the event that all its points are at least a distance $R > 0$ apart. Thus, two parameters need to be estimated, the hard core distance

R and the intensity parameter β. Let us first consider R and write $d_1(\mathbf{x})$ for the minimal distance between points in configuration $\mathbf{x}$. Then, seen as a function of R, $p(\mathbf{x}; (\beta, R))$ is 0 on the interval $(d_1(\mathbf{x}), \infty)$ and increasing on $[0, d_1(\mathbf{x})]$. Hence, $\hat{R} = d_1(\mathbf{x})$ is a maximum likelihood estimator of the hard core distance. For the intensity parameter β on the other hand, no explicit expression for the maximum likelihood estimator seems known [179].

Example 3.18   The *Strauss* or *soft core process* [210] is defined by

$$p(\mathbf{x}; \theta) = \alpha(\theta)\, \beta^{n(\mathbf{x})} \gamma^{s(\mathbf{x})}$$

where $n(\mathbf{x})$ denotes the cardinality of $\mathbf{x}$, $s(\mathbf{x})$ is the number of R−close pairs in $\mathbf{x}$, and $\theta = (\beta, \gamma)$ with $\beta > 0$, $\gamma \in (0,1)$ is the parameter vector of interest.

To find the maximum likelihood equations, note that

$$\alpha(\theta) = \left(\int \beta^{n(\mathbf{x})} \gamma^{s(\mathbf{x})} \, d\pi(\mathbf{x}) \right)^{-1},$$

hence the partial derivatives of $\log p(\mathbf{x}; \theta)$ with respect to β and γ are

$$\frac{n(\mathbf{x})}{\beta} - \frac{1}{\int \beta^{n(\mathbf{x})} \gamma^{s(\mathbf{x})} \, d\pi(\mathbf{x})} \frac{1}{\beta} \int n(\mathbf{x})\, \beta^{n(\mathbf{x})}\, \gamma^{s(\mathbf{x})} \, d\pi(\mathbf{x})$$

and

$$\frac{s(\mathbf{x})}{\gamma} - \frac{1}{\int \beta^{n(\mathbf{x})} \gamma^{s(\mathbf{x})} \, d\pi(\mathbf{x})} \frac{1}{\gamma} \int s(\mathbf{x})\, \beta^{n(\mathbf{x})}\, \gamma^{s(\mathbf{x})} \, d\pi(\mathbf{x})$$

respectively. Consequently, a maximum likelihood estimator solves

$$(n(\mathbf{x}), s(\mathbf{x})) = (\mathbf{E}_\theta n(X), \mathbf{E}_\theta s(X)) \tag{3.12}$$

where $\mathbf{E}_\theta$ denotes the expectation under $p(\cdot; \theta)$. These equations seem hard to solve directly due to the dependence of the expectations $\mathbf{E}_\theta n(X)$ and $\mathbf{E}_\theta s(X)$ on the intractable $\alpha(\theta)$.

The form of (3.12) is typical for many Markov point process models. For example, the maximum likelihood equations for the area-interaction model of example 2.11 are

$$(n(\mathbf{x}), \mu(U_\mathbf{x})) = (\mathbf{E}_\theta n(X), \mathbf{E}_\theta \mu(U_X))$$

where U_X denotes the union of balls around the points of X. In order to derive the general form of the maximum likelihood equations, write [63]

$$p(\mathbf{x}; \theta) = \frac{h_\theta(\mathbf{x})}{Z(\theta)}$$

where $h_\theta(\cdot)$ denotes the unnormalised density and $Z(\theta)$ the reciprocal normalising constant. In statistical physics terminology, $Z(\theta)$ is known as the *partition function* or the *Zustandssumme*. We will assume that all derivatives considered below exist and are sufficiently smooth to allow the order of integration and differentiation to be interchanged. Then, again writing $\mathbf{E}_\theta$ for the expectation with respect to $p(\cdot; \theta)$,

$$-\frac{\partial}{\partial \theta_i} \log Z(\theta) = -\mathbf{E}_\theta \left[\frac{\frac{\partial}{\partial \theta_i} h_\theta(X)}{h_\theta(X)} \right]$$

and hence the maximum likelihood equations are

$$\frac{\frac{\partial}{\partial \theta_i} h_\theta(\mathbf{x})}{h_\theta(\mathbf{x})} = \mathbf{E}_\theta \left[\frac{\frac{\partial}{\partial \theta_i} h_\theta(X)}{h_\theta(X)} \right] \tag{3.13}$$

$i = 1, \ldots, \text{dimension}(\theta)$. In general, a unique solution to (3.13) is not guaranteed.

Example 3.19 Many point process models, including the Strauss and area-interaction processes, form an exponential family [63; 65]

$$p(\mathbf{x}) = \frac{\exp[-\theta' T(\mathbf{x})]}{Z(\theta)} g(\mathbf{x})$$

where $T(\cdot)$ is the canonical sufficient statistic. For these models, the preceding theory is greatly simplified. Indeed, the unnormalised density function is $h_\theta(\mathbf{x}) = g(\mathbf{x}) \exp[-\theta' T(\mathbf{x})]$, so $\frac{\partial}{\partial \theta_i} \log h_\theta(\mathbf{x}) = -T(\mathbf{x})$ does not depend on θ. Therefore (3.13) reduces to

$$-T(\mathbf{x}) = -\mathbf{E}_\theta T(X).$$

Moreover, the matrix of second order derivatives of $\log p(\mathbf{x}; \theta)$ is simply

$$-\text{Cov}_\theta T(X)$$

so that the Fisher information matrix $I(\theta)$ is just $\text{Cov}_\theta T(X)$. Therefore, assuming the covariance matrix of $T(X)$ to be strictly positive definite for all θ, any solution to the maximum likelihood equations is necessarily unique.

The maximum likelihood equations (3.13) cannot be solved directly except in special cases. Nevertheless, the Markov chain Monte Carlo ideas discussed in sections 3.2–3.6 are fruitful in the present context as well [62; 63; 64; 65]. To avoid simulation for all values of θ, we shall take an importance sampling approach [62; 65]. Suppose that a fixed reference point $\psi \in \Theta$ is picked. Then

$$\log \frac{p(\mathbf{x}; \theta)}{p(\mathbf{x}; \psi)} = \log \frac{h_\theta(\mathbf{x})}{h_\psi(\mathbf{x})} - \log \frac{Z(\theta)}{Z(\psi)} \tag{3.14}$$

of which the first term is known in closed form, the second one is not. Moreover, the division by the constant term $p(\mathbf{x}; \psi)$ does not affect the optimisation procedure for θ. Thus, realising that

$$\frac{Z(\theta)}{Z(\psi)} = \mathbf{E}_\psi \left[\frac{h_\theta(X)}{h_\psi(X)} \right],$$

the functional limit theorem for Monte Carlo samplers discussed in the previous sections implies that the log likelihood ratio function (3.14) can be estimated by

$$l_n(\theta) = \log \frac{h_\theta(\mathbf{x})}{h_\psi(\mathbf{x})} - \log \left(\frac{1}{n} \sum_{i=1}^n \frac{h_\theta(X_i)}{h_\psi(X_i)} \right) \tag{3.15}$$

where $X_1, \ldots, X_n$ are samples from $p(\cdot; \psi)$.

Example 3.20 For the Strauss process of example 3.18, the Monte Carlo log likelihood ratio with respect to $\psi = (\beta_0, \gamma_0)$ is

$$n(\mathbf{x}) \log \left(\frac{\beta}{\beta_0} \right) + s(\mathbf{x}) \log \left(\frac{\gamma}{\gamma_0} \right) - \log \left[\frac{1}{n} \sum_{i=1}^n \left(\frac{\beta}{\beta_0} \right)^{n(X_i)} \left(\frac{\gamma}{\gamma_0} \right)^{s(X_i)} \right]$$

where $X_1, \ldots, X_n$ are realisations of a Strauss process with parameter vector ψ.

The importance sampling estimator (3.15) performs well for θ-values not too far from the reference value ψ. If the parameter space is large, a practical approach is to partition, take a ψ value for each region and use interpolation to compute $l_n(\theta)$.

Finally, having computed $l_n(\theta)$ as in (3.15), given an observed pattern $X = \mathbf{x}$, an approximate maximum likelihood solution can be found by solving

$$\frac{\frac{\partial}{\partial \theta} h_\theta(\mathbf{x})}{h_\theta(\mathbf{x})} = \frac{\sum_{i=1}^{n} \frac{\frac{\partial}{\partial \theta} h_\theta(X_i)}{h_\psi(X_i)}}{\sum_{i=1}^{n} \frac{h_\theta(X_i)}{h_\psi(X_i)}},$$

the sampling analogue of (3.13).

3.8 Estimation based on the conditional intensity

In the subsections below we will discuss in some detail two alternative techniques that are based on the Papangelou conditional intensity introduced in chapter 1.

3.8.1 *Takacs–Fiksel estimation*

The *Takacs–Fiksel method* [53; 54; 212; 213; 214] is based on equation (1.16), which, combined with the assumption that the point process X is stationary, leads to the *Nguyen–Zessin identity*[152]

$$\lambda \, \mathbf{E}_y^! g(X) = \mathbf{E}\left[\lambda(y; X) \, g(X)\right] \qquad (3.16)$$

holding for any non-negative, measurable function $g : N^{\mathrm{lf}} \to \mathbb{R}^+$ and any stationary Markov point process X on $\mathbb{R}^d$ with finite intensity λ. The expectation on the left hand side of (3.16) is with respect to the reduced Palm distribution of X at some $y \in \mathbb{R}^d$, while $\lambda(y; \mathbf{x})$ is the Papangelou conditional intensity of X at y (see chapter 1 as well as [67; 68; 110; 132; 98] for further details). Although in practice a point pattern is observed only within a bounded window $\mathcal{X}$ and hence cannot be stationary, this fact is mostly conveniently ignored and the data pattern is treated as if it were a partial observation from a stationary pattern extended over the whole space. A word of caution is needed though, since not every point process – even if its clique interaction function is translation invariant – can be extended uniquely to a stationary process on $\mathbb{R}^d$ [165; 191].

The next examples present some commonly used functional forms for $g(\cdot)$.

Example 3.21 By definition 1.1, $g(X) = N(B(0, R))$, the number of points falling in a ball of radius $R > 0$, is a random variable. In this case, the left hand side of (3.16) is proportional to the *second reduced moment function* $K(\cdot)$ [174],

$$\lambda \, \mathbf{E}_0^! n(X \cap B(0, R)) = \lambda^2 K(R).$$

Example 3.22 Set

$$g(X) = \frac{\mathbb{I}\{X \cap B(0, R) = \emptyset\}}{\lambda(0; X)}.$$

Then $\mathbf{E}\,[\lambda(0; X)g(X)] = 1 - F(R)$, where $F(\cdot)$ is the empty space function [46] (cf. section 1.8). If X is Markov at range r, for $R > r$ the left hand side of (3.16) reduces to

$$\lambda \, \mathbf{E}_0^! \left[\frac{\mathbb{I}\{X \cap B(0, R) = \emptyset\}}{\lambda(0; X)} \right] = \lambda \, \mathbf{E}_0^! \left[\frac{\mathbb{I}\{X \cap B(0, R) = \emptyset\}}{\lambda(0; \emptyset)} \right]$$

$$= \frac{\lambda}{\lambda(0; \emptyset)}(1 - G(R))$$

where $G(\cdot)$ is the distribution function of inter-event distances [46]. The ratio $(1 - G(R))/(1 - F(R))$ is the J–statistic [122] for measuring the strength and range of interactions in spatial patterns (see section 1.8).

Example 3.23 A convenient choice is $g(X) = N(B(0, R))/\lambda(0; X)$ as suggested by Fiksel [53; 54]. Then the right hand side of (3.16) is constant.

The Nguyen–Zessin formula can be used to estimate model parameters. The idea is to choose functions $g_i(\cdot)$, $i = 1, \ldots, m$, one for each component of the parameter vector θ, and to estimate both sides in (1.16). Solving the resulting set of equations (in a least squares sense if necessary) yields estimates for the model parameters. More precisely, writing $\mathbf{x} - x$ for the point pattern **x** translated over the vector $x \in \mathbb{R}^d$, the left hand side of (3.16) can be estimated by

$$\widehat{L_i(\mathbf{x}; \theta)} = \frac{1}{\mu(\mathcal{X})} \sum_{x \in \mathbf{x}} g_i((\mathbf{x} - x) \setminus \{0\}) \qquad (3.17)$$

with $\mu(\mathcal{X})$ denoting the volume of window $\mathcal{X}$; an estimator for the right hand side is

$$\widehat{R_i(\mathbf{x}; \theta)} = \frac{1}{L} \sum_{j=1}^{L} \lambda(y_j; \mathbf{x}) \, g_i(\mathbf{x} - y_j) \qquad (3.18)$$

where $\{y_1, \ldots, y_L\} \subseteq \mathcal{X}$ is a fixed collection of points. Thus, $\widehat{L_i(\mathbf{x}; \theta)}$ examines the pattern from the point of view of each data point $x \in \mathbf{x}$, whereas $\widehat{R_i(\mathbf{x}; \theta)}$ is based on arbitrary placed centres y_j. Finally, a Takacs–Fiksel estimator [40; 48; 53; 54; 208; 196; 212; 213; 214]. is obtained by minimising

$$\sum_{i=1}^{m} \{\widehat{L_i(\mathbf{x}; \theta)} - \widehat{R_i(\mathbf{x}; \theta)}\}^2 \qquad (3.19)$$

with respect to $\theta = (\theta_1, \ldots, \theta_m)$.

Ignoring edge effects, by the Campbell–Mecke formula (1.15)

$$\mathbf{E}\widehat{L_i(X; \theta)} = \frac{\lambda}{\mu(\mathcal{X})} \int_{\mathcal{X}} \int_{N^{lf}} g_i(X - x) \, dP_x^!(X) \, dx = \lambda \, \mathbf{E}_0^! g_i(X)$$

where we have used the fact that X is stationary in the last equality. Moreover,

$$\mathbf{E}\widehat{R_i(X; \theta)} = \frac{1}{L} \sum_{j=1}^{L} \mathbf{E}\left[\lambda(0; X - y_j) \, g_i(X - y_j)\right] = \mathbf{E}\left[\lambda(0; X) \, g_i(X)\right].$$

Hence both $\widehat{L_i(X; \theta)}$ and $\widehat{R_i(X; \theta)}$ are unbiased. Expressions for the variance of these estimators can be computed in a similar fashion. However, $\widehat{L_i}$ and $\widehat{R_i}$ are *not* necessarily unbiased if the point process X is replaced by its restriction to some observation window $\mathcal{X}$. Several strategies are available to deal with this problem. For instance, statistics such as $F(R)$, $G(R)$, $J(R)$ and $K(R)$ can be estimated by taking into account only those points that are further than R away from the window boundary, a technique called *minus sampling* [179; 208]. More sophisticated techniques are based on various edge correction terms [5; 35; 84; 179; 208].

Example 3.24 Further to example 3.22, a naive estimator of the empty space function $F(R)$ is simply the fraction of test points y_j for which the ball $B(y_j, R)$ contains points of $\mathbf{x}$. The minus sampling estimator is obtained by ensuring all points y_j are at least a distance R away from the boundary of $\mathcal{X}$. This may be inefficient if R is large; indeed if the distance $d(y_j, \mathbf{x})$ from y_j to the closest point of $\mathbf{x}$ is less than the distance $d(y_j, \mathcal{X}^c)$ from y_j to the boundary of $\mathcal{X}$, y_j may be retained without introducing bias. This observation underlies the Hanisch-style estimator [35; 83]

$$\widehat{F(R)} = \sum_{r \leq R} \left[\frac{n(\{j : d(y_j, \mathbf{x}) = r \leq d(y, \mathcal{X}^c)\})}{n(\{j : d(y_j, \mathcal{X}^c) \geq r\})} \right]$$

with $n(\cdot)$ as usual denoting cardinality. Both the minus sampling and Hanisch-style estimator are unbiased estimators of $F(R)$.

For non-homogeneous models, the Takacs–Fiksel idea still applies, but the more general formula (1.16)

$$\mathbf{E}\left[\sum_{x \in X} g(x, X \setminus \{x\}) \right] = \mathbf{E}\left[\int g(x, X)\, \lambda(x; X)\, dx \right]$$

must be used. The left hand side can be estimated by $\widehat{L(\mathbf{x}; \theta)} = \sum_{x \in \mathbf{x}} g(x, \mathbf{x} \setminus \{x\})$. An estimator for the right hand side is $\widehat{R(\mathbf{x}; \theta)} = \frac{\mu(\mathcal{X})}{L} \sum_{j=1}^{L} \lambda(y_j; \mathbf{x})$ $g(y_j, \mathbf{x})$ where $\{y_1, \ldots, y_L\} \subseteq \mathcal{X}$ is a collection of independent, uniformly distributed test points. Taking a function $g_i(\cdot)$ for each component θ_i of the parameter vector θ, and minimising (3.19) with respect to θ completes the procedure.

3.8.2 *Maximum pseudo-likelihood estimation*

In Markov point process models, the density $p(\mathbf{x}; \theta)$ generally speaking is specified up to an intractable normalising constant $\alpha(\theta)$ only; the conditional intensity on the other hand tends to be easy to compute. Thus, pseudo-likelihood based estimation aims to optimise*

$$PL(\mathbf{x};\theta) = \exp\left[-\int \lambda_\theta(u;\mathbf{x})\,d\nu(u)\right]\prod_{x\in\mathbf{x}}\lambda_\theta(x;\mathbf{x}) \qquad (3.20)$$

which avoids computation of $\alpha(\theta)$. For Poisson processes, the conditional intensity is equal to the intensity function and hence maximum pseudo-likelihood estimation is equivalent to maximising the likelihood (cf. section 3.7). In general, $PL(\cdot;\theta)$ is only an approximation of the true likelihood. However, no sampling is needed, and the computational load will be considerably smaller than that of the likelihood based approach [13; 20; 24; 95; 96; 97; 69; 196; 214]. In case of strong interaction, maximum pseudo-likelihood estimation may be followed by one or a few steps of the Newton-Raphson algorithm for optimising the (proper) likelihood [92].

Example 3.25 For the Strauss process of example 3.18, the conditional intensity is $\lambda(u;\mathbf{x}) = \beta\,\gamma^{s(u;\mathbf{x})}$, where $s(u;\mathbf{x})$ denotes the number of points in $x\in\mathbf{x}$ with $0 < d(u,x) \le R$. Hence the pseudo-likelihood equals

$$PL(\mathbf{x};\beta,\gamma) = \exp\left[-\int_\chi \beta\,\gamma^{s(u;\mathbf{x})}\,d\nu(u)\right]\prod_{x\in\mathbf{x}}\left(\beta\,\gamma^{s(x;\mathbf{x})}\right)$$

and maximum pseudo-likelihood estimates of β and γ solve

$$n(\mathbf{x}) = \beta\int_\chi \gamma^{s(u;\mathbf{x})}\,d\nu(u)$$

$$\sum_{x\in\mathbf{x}} s(x;\mathbf{x}) = \beta\int s(u;\mathbf{x})\,\gamma^{s(u;\mathbf{x})}\,d\nu(u).$$

The form of these equations is typical for exponential families (cf. example 3.19).

Maximum pseudo-likelihood estimation often is just a special case of the Takacs–Fiksel method. To see this, suppose that the conditional intensity and the partial derivatives of $\log \lambda_\theta(u;\mathbf{x})$ with respect to the model parameters exist and are sufficiently smooth to allow the order of integration and

*Some authors prefer to condition on the number of points and use $\prod_{x\in\mathbf{x}}\left\{\frac{\lambda_\theta(x;\mathbf{x}\setminus\{x\})}{\int_\chi \lambda_\theta(\xi;\mathbf{x}\setminus\{x\})\,d\nu(\xi)}\right\}$ instead of $PL(\mathbf{x};\theta)$.

differentiation to be interchanged. Set

$$g_i(u, X) = \frac{\partial}{\partial \theta_i} \log \lambda_\theta(u; X).$$

Substitution into (1.16) yields

$$\mathbf{E} \left[\sum_{x \in \mathbf{x}} \frac{\partial}{\partial \theta_i} \log \lambda_\theta(x; X) \right] = \mathbf{E} \left[\int_{\mathcal{X}} \lambda_\theta(u; X) \frac{\partial}{\partial \theta_i} \log \lambda_\theta(u; X) \, d\nu(u) \right]$$

using the fact that $\lambda_\theta(x; X) = \lambda_\theta(x; X \setminus \{x\})$ by definition. Equivalently,

$$\mathbf{E} \left[\frac{\partial}{\partial \theta} \log PL(X; \theta) \right] = 0$$

from which we obtain the maximum pseudo-likelihood equations upon replacing the expectation by its empirical counterpart.

3.9　Goodness of fit testing

The previous sections were concerned with the estimation of model parameters. Having done so, the next step may be to assess the goodness of fit. Thus, let $\{P_\theta : \theta \in \Theta\}$ be a family of probability distributions parameterised by a vector θ in some space Θ, and suppose we want to test the null hypothesis $R(\theta) = 0$ for some differentiable constraint function $R : \Theta \mapsto \mathbb{R}^d$ against the alternative $R(\theta) \neq 0$. Then the Wald test statistic W is given by [63]

$$W = R(\hat{\theta})^T \left(R'(\hat{\theta})^T I(\hat{\theta})^{-1} R'(\hat{\theta}) \right)^{-1} R(\hat{\theta})$$

where $\hat{\theta}$ is the maximum likelihood estimator under the alternative hypothesis, the superscript T denotes transposition, and $I(\cdot)$ is the Fisher information matrix. Since (Monte Carlo estimates of) the maximum likelihood estimator and Fisher information matrix are already available if we come to this stage, computation of W is straightforward.

Example 3.26　Consider the exponential family of point process densities

$$p(\mathbf{x}) = \frac{\exp[-\theta' T(\mathbf{x})]}{Z(\theta)} g(\mathbf{x})$$

studied in example 3.19, and suppose the parameter vector $\theta = (\theta_1, \theta_2) \in \mathbb{R}^2$ consists of an intensity parameter θ_1 and an interaction parameter θ_2. By the theory outlined in section 3.7, $\hat{\theta}$ can be approximated by optimising the Monte Carlo log likelihood ratio function (3.15); an estimate for the Fisher information function is the Monte Carlo covariance matrix of the canonical sufficient statistic $T(X)$.

The hypothesis that no further interaction is imposed on the base model $g(\mathbf{x})$ is represented by the constraint $\theta_2 = 0$. Thus, $R((\theta_1, \theta_2)) = \theta_2$ is the projection on the second coordinate and the Wald test statistic is simply $\hat{\theta}_2^2 / I(\hat{\theta})_{22}^{-1}$ where $I(\hat{\theta})_{22}^{-1}$ is the lower right coordinate of the inverse Fisher information matrix at the maximum likelihood estimator $\hat{\theta}$. Since the difference in dimension between the alternative and null hypotheses equals 1, W should be compared to the quantiles of a χ^2 distribution with one degree of freedom [63].

For arbitrary null and alternative hypotheses, a likelihood ratio test statistic may be used. However, since maximum likelihood estimators for both hypotheses are involved, extra sampling is required to perform the test [63].

In a non-parametric approach, test statistics may be based on the empty space function, the nearest-neighbour distance distribution function, and the J-statistic (section 1.8), or Ripley's K-function (section 3.8). More recently, granulometries [120; 179] and Euler characteristic based test statistics [136] have been advocated. Little is known about the distribution of any of these statistics but sampling under a null distribution model is relatively easy, thus allowing a Monte Carlo approach.

More specifically, consider the problem of investigating whether a particular model P_0 applies, that is to test $\mathcal{H}_0 : X \sim P_0$ against the alternative $\mathcal{H}_1 : X \not\sim P_0$. A Monte Carlo test [21; 45] for this problem can be designed as follows.

Algorithm 3.6 Given a summary statistic $V(\cdot)$,

- sample $n - 1$ independent realisations $\mathbf{x}_1, \ldots, \mathbf{x}_{n-1}$ from the null hypothesis distribution P_0;
- compute $V_1 = V(\mathbf{x}_1), \ldots, V_{n-1} = V(\mathbf{x}_{n-1})$;
- reject the null hypothesis if $V(\mathbf{x})$ is amongst the k extremes where

x is the observation pattern (in case of ties, take the least extreme rank for the data).

The above specifies an exact test of size k/n.

In the literature, one often encounters plots such as Figure 1.12 of an empirical V–statistic together with envelopes based on say $n - 1 = 19$ simulations. It is important to note however, that the test statistic must be fixed in advance, not after inspection of the picture. Thus, if a single value statistic $V = \widehat{J(r_0)}$ (say) is used, then r_0 has to be prespecified. A fairer comparison may be to take the integrated quadratic difference

$$V_i = \int_0^{r_0} \{\widehat{J_i(r)} - J(r)\}^2 dr$$

or the supremum statistic

$$V_i = \sup_{r \leq r_0} \{\widehat{J_i(r)} - J(r)\}^2$$

$(i = 0, \ldots, n - 1)$ where $\widehat{J_i(\cdot)}$ is computed from $\mathbf{x}_i$ and $J(\cdot)$ denotes the (theoretical) J–statistic under the null hypothesis.

Similar ideas underly the least squares estimation method. Suppose, a family of point process densities $p(\cdot; \theta)$ is given, parameterised by $\theta \in \Theta$. If e.g. the J–statistic is used, write $J(r, \theta)$ for the theoretical J–function under the parameter value θ, and as before $\widehat{J(r)}$ for an estimate based on the data. Setting

$$D(\theta) = \int_0^{r_0} \{\widehat{J(r)} - J(r, \theta)\}^2 dr,$$

θ may be estimated by

$$\hat{\theta} = \mathrm{argmin}_\theta D(\theta).$$

If the theoretical value $J(r, \theta)$ is not known, it may be replaced by an estimate based on (independent) Monte Carlo samples. Although the numerical optimisation problem can be performed using standard least squares software, a drawback of the method is that virtually nothing is known about the bias, variance and other statistical properties of the resulting estimator $\hat{\theta}$.

3.10 Discussion

In this section we considered sampling and inference for spatial point patterns. For simplicity, we have used simple unmarked point processes throughout. However, since a marked point process is a point process (on $\mathcal{X} \times \mathcal{K}$), the techniques presented can easily be adapted to the marked case.

Sampling can be carried out by a spatial birth-and-death process as in the unmarked case, and, provided the conditional intensity is uniformly bounded from above, the process forms the basis for a coupling from the past algorithm.

The Metropolis–Hastings algorithm requires a bit more care. Of course, algorithm 3.1 may be used, but better mixing may be obtained by allowing moves that change the marks. For instance, if the mark describes the shape of an object, it is more efficient to slightly change one or more shape descriptors than to remove the marked point (resulting in a low probability state) and subsequently replace it by a new point with other shape parameters [6; 116]. Moreover, especially in biological applications, split and merge strategies (example 3.4) are useful [74; 189; 190].

Likelihood based inference as described in section 3.7 carries over in a straightforward fashion to marked point processes. The pseudo-likelihood at a configuration $\mathbf{y} \subseteq \mathcal{X} \times \mathcal{K}$ is

$$PL(\mathbf{y}; \theta) = \exp\left[-\int_\mathcal{X} \int_\mathcal{K} \lambda_\theta((u,k); \mathbf{y}) \, d\nu(u) \, dm(k)\right] \prod_{i=1}^n \lambda_\theta((u,k); \mathbf{y})$$

where $\lambda_\theta(\cdot; \cdot)$ is the conditional intensity, parameterised by θ. As in the unmarked case, $PL(\mathbf{y}; \theta)$ is known explicitly, allowing for numerical optimisation over θ [69].

More generally, the Takacs–Fiksel estimation method uses the Nguyen–Zessin equality for Markov marked point processes

$$\mathbf{E}\left[\sum_{(u,k) \in Y} g((u,k), Y \setminus \{(u,k)\})\right] =$$

$$\mathbf{E}\left[\int_\mathcal{X} \int_\mathcal{K} g((u,k), Y) \, \lambda_\theta((u,k); Y) \, d\nu(u) \, dm(k)\right]$$

holding for any non-negative measurable function $g(\cdot, \cdot)$. If the parameter vector θ has m components, one chooses m functions $g_i(\cdot, \cdot)$, plugs in estimates of the left and right hand side of the Nguyen–Zessin identity (as an average over the points of the process, respectively a window average) and solves the resulting set of equations. Pseudo–likelihood estimation is the special case $g_i((x, k); \mathbf{y}) = \frac{\partial}{\partial \theta_i} \log \lambda_\theta((x, k); \mathbf{y})$.

Finally, although we have concentrated on parametric inference, some authors have studied non-parametric estimation of the pair interaction function [49; 90] or the first-order intensity function [50].

Chapter 4

Applications

4.1 Modelling spatial patterns

In the previous chapters we already encountered a range of Markov point process models. Just to mention a few, the class of pairwise interaction densities offers a myriad of models for repulsive patterns, while the area-interaction model and the cluster processes of example 1.14 provide alternatives for more aggregated patterns. In this chapter we will introduce some new models and treat old ones in more detail.

There are several ways of specifying a Markov point process distribution. Naturally, one may define an (unnormalised) density, but it is often more convenient to think locally. Indeed the most important property of a Markov density $p(\cdot)$ is the fact that the conditional intensity

$$\lambda(u; \mathbf{x}) = \frac{p(\mathbf{x} \cup \{u\})}{p(\mathbf{x})}, \qquad u \notin \mathbf{x},$$

depends only on u and its neighbours in $\mathbf{x}$. Assuming $p(\cdot) > 0$, knowing the conditional intensity is sufficient to derive the full probability distribution by simply adding the points of $\mathbf{x}$ one at a time. More surprisingly, specifying a functional form for the conditional intensity is often sufficient (see sections 4.2 and 4.3). Another attractive route is to take advantage of the Hammersley–Clifford theorem 2.2 and define the clique interaction function. Some care must be taken though to ensure that the proposal leads to a proper density that integrates to unity. Sufficient conditions are that the interaction function is less than or equal to 1, in which case the model is said to be *purely inhibitive* [24], or that the number of points almost

115

surely does not exceed some $n \in \mathbb{N}$ as for the hard core process. In more set-geometric models such as the area-interaction process, uniform bounds on the contribution of each of the balls (or other grains in a more general set-up, cf. example 1.12) can usually be found which imply the integrability of the model.

Throughout this chapter, unless stated otherwise, we will assume that X is a Markov point process on a complete, separable metric space $(\mathcal{X}, d)$. The reference distribution is that of a Poisson process with finite, non-atomic intensity measure $\nu(\cdot)$.

4.2 Pairwise interaction processes

In this section we will assume that the interaction between points is restricted to pairs [46; 176; 179], i.e. the density $p(\cdot)$ can be written as

$$p(\mathbf{x}) = \alpha \prod_{x \in \mathbf{x}} \beta(x) \prod_{u,v \in \mathbf{x}:u \sim v} \gamma(u, v) \tag{4.1}$$

where $\alpha > 0$ is the normalising constant, $\beta : \mathcal{X} \to \mathbb{R}^+$ is a Borel measurable intensity function and $\gamma : \mathcal{X} \times \mathcal{X} \to \mathbb{R}^+$ is a symmetric, Borel measurable interaction function.

One of the best known examples of (4.1), proposed by Strauss [210] for the apparent clustering of redwood seedlings around older stumps, is the soft core process with density

$$p(\mathbf{x}) = \alpha \, \beta^{n(\mathbf{x})} \, \gamma^{s(\mathbf{x})} \tag{4.2}$$

where $\beta, \gamma > 0$, $n(\mathbf{x})$ denotes the number of points in $\mathbf{x}$, and $s(\mathbf{x})$ is the number of neighbour pairs $u, v \in \mathbf{x}$ with $d(u, v) \leq R$ for some $R > 0$. Hence the only non-trivial clique interaction terms are β for singletons and γ for doublets. Several special cases deserve mentioning: $\gamma = 0$ corresponds to the hard core process studied in chapter 3 (see Figure 2.1), $\gamma = 1$ to a Poisson process of rate β.

First consider parameter values $\gamma \in (0, 1)$. Then the Strauss process is locally stable as

$$\lambda(u; \mathbf{x}) = \beta \, \gamma^{s(\mathbf{x} \cup \{u\}) - s(\mathbf{x})} \leq \beta \tag{4.3}$$

is uniformly bounded. Since $s(\mathbf{x} \cup \{u\}) - s(\mathbf{x})$ equals the number of points in $\mathbf{x}$ that are closer than R to the new point u, the model is Markov at range R. It tends to favour configurations with fewer R-close pairs than the reference Poisson process. A typical realisation in the unit square is shown in Figure 4.1. We used algorithm 3.4 to sample a pattern in the enlarged square $[-R, 1 + R]^2$, and then truncated to $[0, 1]^2$ to avoid edge effects, cf. theorem 2.1. An alternative is the auxiliary variables algorithm (example 3.12), implemented by a transition function [140] based on an adaptive sequence of conditional Poisson processes. Suppose the current state is $\mathbf{x}$ and let W_1 be a Poisson process of rate β. For $i = 1, 2, \ldots$, let W_{i+1} be a Poisson process restricted to the event $\{\prod_{u \sim v \in W_i} \gamma(u, v) \leq \prod_{u \sim v \in W_{i+1}} \gamma(u, v)\} = \{s(W_{i+1}) \leq s(W_i)\}$. Given a random variable U that is uniformly distributed on the unit interval, the algorithm jumps to the first W_i for which $\prod_{u \sim v \in W_i} \gamma(u, v)$ exceeds $U \prod_{u \sim v \in \mathbf{x}} \gamma(u, v)$. Clearly this transition function is monotone in the number of neighbour pairs (and in $\prod_{u \sim v} \gamma(u, v)$ for more general transition functions that are bounded by 1), hence amenable to coupling from the past. The algorithm terminates almost surely, due the fact that the empty set is an atom of the Poisson process. We prefer the Propp–Wilson algorithm as it does not require adaptive rejection sampling.

Unfortunately, although intended to model clustering of seedlings, (4.2) is not-integrable for $\gamma > 1$, unless one restricts attention to configurations with a fixed number of points [101]. Summarising, we have the following result [101; 210].

Theorem 4.1   *The Strauss process is well-defined for all $\beta > 0$ and $\gamma \in [0, 1]$. It is Markov at range R with interaction function $\gamma(u, v) = \gamma \, \mathbb{I}\{d(u, v) \leq R\}$. If there exists a non-finite Borel set $A \subseteq \mathcal{X}$ such that $\nu(A) > 0$ and $d(u, v) \leq R$ for all $u, v \in A$, $p(\cdot)$ cannot be normalised to integrate to unity.*

Proof.   To show that the Strauss process is well-defined for $\gamma \leq 1$, note that in this case $p(\mathbf{x}) \leq \alpha \beta^{n(\mathbf{x})}$. The right hand side is proportional to a density of a Poisson process with intensity β, hence integrable. It follows that a value of α can be chosen so that $p(\cdot)$ integrates to unity. The Markov property follows from (4.3).

Next, consider the case $\gamma > 1$. We will try and find patterns with order n^2 interactions between n points. To do so, let $A \subseteq \mathcal{X}$ be a Borel set such that $\nu(A) > 0$ and $d(u, v) \leq R$ for all $u, v \in A$. Such a set exists by

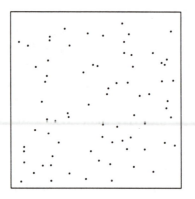

Fig. 4.1   Realisation of a Strauss model in the unit square with intensity parameter $\beta = 100$ and interaction parameter $\gamma = 0.5$. The interaction range is 0.05.

assumption. Then for all configurations $\mathbf{x} \subseteq A$ consisting of n points, the number of neighbour pairs is $s(\mathbf{x}) = \frac{1}{2} n(n-1)$. Therefore, writing $\pi(\cdot)$ for the distribution of the reference Poisson process,

$$
\begin{aligned}
\int_{N^f} p(\mathbf{x}) \, d\pi(\mathbf{x}) &= \sum_{n=0}^{\infty} \frac{e^{-\nu(\mathcal{X})}}{n!} \int_{\mathcal{X}} \cdots \int_{\mathcal{X}} p(\{x_1, \ldots, x_n\}) \, d\nu(x_1) \cdots d\nu(x_n) \\
&\geq \sum_{n=1}^{\infty} \frac{e^{-\nu(\mathcal{X})}}{n!} \, \alpha \, \beta^n \, \gamma^{n(n-1)/2} \, \nu(A)^n = \infty
\end{aligned}
$$

unless $\alpha = 0$. It follows that the Strauss model with $\gamma > 1$ is ill-defined. $\square$

It should be noted that theorem 4.1 remains valid for neighbourhood relations $\sim$ other than the fixed range relation (example 2.1).

Conditional on a total number of n points, the joint density for the locations is

$$
j_n(x_1, \ldots, x_n) \propto \gamma^{s(\{x_1, \ldots, x_n\})}
$$

with respect to the product measure $\nu(\cdot)^n$ on $\mathcal{X}^n$. Clearly $j_n(\cdot,\dots,\cdot)$ is well-defined for $\gamma > 1$ as well, and in this case will result in clustered realisations with high probability.

Above, the Strauss model was defined by means of a clique interaction term γ for each pair of neighbouring points. Alternatively, we could have tried and define the model in terms of the conditional intensity. Indeed, if we had simply asked for $\lambda(u; \mathbf{x})$ to be a function of the number of neighbours $s(u, \mathbf{x})$ of the point $u \in \mathcal{X}$ in $\mathbf{x}$, under mild conditions the Strauss model turns out to be the only choice [77; 101; 210].

Theorem 4.2 *Suppose there exist $v \sim w$ in $\mathcal{X}$ and a series $x_1, x_2, \dots \in \mathcal{X}$ such that $v \sim x_i$ but $w \not\sim x_i$ for all $i = 1, 2, \dots$. For $u \in \mathcal{X}$, write $s(u, \mathbf{x})$ for the number of neighbours of u in $\mathbf{x}$. Then the Strauss density (4.2) is the only hereditary function $p : N^{\mathrm{f}} \to \mathbb{R}^+$ for which*

$$\frac{p(\mathbf{x} \cup \{u\})}{p(\mathbf{x})} = g(s(u, \mathbf{x}))$$

for some function $g : \mathbb{N}_0 \to [0, \infty)$.

For the fixed range relation, a sufficient condition for the characterisation to hold is that $\mathcal{X}$ contains a sufficiently large ball.

Proof. The Strauss density (4.2) is hereditary. Moreover, $\lambda(u; \mathbf{x}) = \beta \gamma^{s(u,\mathbf{x})}$, so $g(s(u, \mathbf{x}))$ has the required form.

Reversely, suppose the ratio $\frac{p(\mathbf{x} \cup \{u\})}{p(\mathbf{x})}$ is a function of the number of neighbour pairs. Let $\alpha = p(\emptyset)$ and $\beta = g(0)$. Firstly, if $\alpha = 0$, then since $p(\cdot)$ is hereditary, $p(\cdot) \equiv 0$ and we can arbitrarily set $\gamma = 1$. Secondly, if $\alpha > 0$ but $\beta = 0$, or equivalently $p(\emptyset) > 0$ and $g(0) = 0$, then $p(\{u\}) = g(0) = 0$ for all singleton sets. Since $p(\cdot)$ is hereditary by assumption, $p(\mathbf{x}) = 0$ except for $\mathbf{x} = \emptyset$ and again we can take $\gamma = 1$.

Finally, consider the case that both α and β are strictly positive. Set $\gamma = g(1)/g(0)$. Now, if $\gamma = 0$ then $p(\mathbf{x}) = 0$ whenever $s(\mathbf{x}) > 0$, a hard core process. It remains to consider the case $\gamma > 0$. Define, for $n \in \mathbb{N}$, $\mathbf{x}_n = \{x_1, \dots, x_n\}$ and $\mathbf{y}_n = \mathbf{x}_n \cup \{v, w\}$. Adding v and w to $\mathbf{x}_n$ in both possible orders, we obtain

$$p(\mathbf{y}_n) = p(\mathbf{x}_n)\, g(0)\, g(n+1) = p(\mathbf{x}_n)\, g(n)\, g(1). \tag{4.4}$$

We will show by induction that $g(m) = \beta \gamma^m$. For $m \in \{0, 1\}$ it is clear that $g(m) = \beta \gamma^m$, and for any $\mathbf{z}$ with $n(\mathbf{z}) \le m + 1$, $p(\mathbf{z}) = \alpha \beta^{n(\mathbf{z})} \gamma^{s(\mathbf{z})} > 0$. Thus, suppose $p(\mathbf{x}_m) > 0$ and $g(n) = \beta \gamma^n$ for all $n \le m$. By (4.4) with $n = m$

$$g(m + 1) = \frac{g(m) \, g(1)}{g(0)} = \beta \gamma^{m+1} > 0$$

and $p(\mathbf{y}_m) > 0$. More generally for any $\mathbf{z}$ with at most $m + 2$ members, $p(\mathbf{z})$ involves only $g(0), \ldots g(m + 1)$ hence $p(\mathbf{z}) > 0$. We conclude that $p(\cdot) > 0$ and $g(n) = \beta \gamma^n$ for any $n = 0, 1, \ldots$. Consequently, for $\mathbf{z} = \{z_1, \ldots, z_n\}$

$$
\begin{aligned}
p(\mathbf{z}) &= p(\emptyset) \frac{p(\{z_1\})}{p(\emptyset)} \cdots \frac{p(\mathbf{z})}{p(\mathbf{z} \setminus \{z_n\})} \\
&= \alpha \prod_{i=1}^{n} g(s(z_i, \{z_1, \ldots, z_{i-1}\})) = \alpha \beta^n \gamma^{s(\mathbf{z})}
\end{aligned}
$$

as required.   □

The general form of (4.1) offers a variety of models. The intensity function $\beta(\cdot)$ can be used to model spatial inhomogeneity or favour certain points over others. The interaction terms $\gamma(u, v)$ often are a function of the distance between u and v.

Example 4.1 Quadratically decaying interaction can be modelled by [48]

$$\gamma(u, v) = \begin{cases} 1 - (1 - d(u, v)^2/R^2)^2 & \text{if } d(u, v) \le R \\ 1 & \text{if } d(u, v) > R \end{cases} . \quad (4.5)$$

In words, R-close points repel each other with a strength depending on the distance between them. A graph of the function $r \mapsto 1 - (1 - r^2/R^2)^2$ for $R = 0.2$ is plotted in Figure 4.2.

Alternatively [49], one may take a linearly decaying interaction function $\gamma(u, v) = \frac{d(u,v)}{R} \, \mathrm{I}\{d(u, v) \le R\}$, but a disadvantage in comparison to (4.5) is that the derivative is not continuous.

Note that (4.5) is a symmetric, Borel measurable function on $\mathcal{X} \times \mathcal{X}$ that is bounded by 1, so provided the intensity function $\beta(\cdot)$ is integrable,

$$\int_{N^f} p(\mathbf{x}) \, d\pi(\mathbf{x}) \le \alpha \sum_{n=0}^{\infty} \frac{e^{-\nu(\mathcal{X})}}{n!} \left(\int_{\mathcal{X}} \beta(u) \, d\nu(u) \right)^n < \infty$$

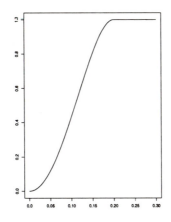

Fig. 4.2 The graph of $\gamma(r) = 1 - (1 - r^2/0.04)^2$ (vertical axis) against r (horizontal axis).

where $\pi(\cdot)$ is the distribution of the reference Poisson process. Hence (4.1) is a well-defined point process density.

Example 4.2 A log-polynomially decaying interaction function such as

$$\gamma(u,v) = 1 - \exp\left[-\left(\frac{d(u,v)}{\theta}\right)^2,\right] \tag{4.6}$$

$\theta > 0$, was proposed in [155], see Figure 4.3. Note that although $\gamma(u,v) \to 1$ as $d(u,v) \to \infty$, points arbitrarily far apart interact.

Example 4.3 The *Lennard–Jones interaction function* on a compact set $\mathcal{X} \subset \mathbb{R}^d$ is defined as

$$\gamma(u,v) = \exp\left[\alpha\left(\frac{\sigma}{d(u,v)}\right)^6 - \beta\left(\frac{\sigma}{d(u,v)}\right)^{12}\right] \tag{4.7}$$

for $\alpha \geq 0$, $\beta, \sigma > 0$. This model for interacting particles in a liquid or dense gas [25; 80; 191] combines close range repulsion with clustering at larger scales. As (4.6), $\gamma(u,v) \to 1$ as $d(u,v) \to \infty$ but seen as a function

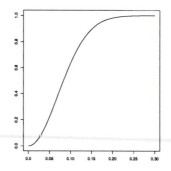

Fig. 4.3   The graph of $\gamma(r) = 1 - \exp\left[-\left(\frac{r}{0.1}\right)^2\right]$ (vertical axis) against r (horizontal axis).

on $\mathbb{R}^d$, (4.7) exhibits an infinite range of interaction. For a plot of $r \mapsto \alpha\left(\frac{\sigma}{r}\right)^6 - \beta\left(\frac{\sigma}{r}\right)^1 2$ with $\alpha = 100$, $\beta = 1$ and $\sigma = 0.2$, see Figure 4.4.

Example 4.4   The Saunders–Kryscio–Funk [193] model joins the Strauss interaction function with a hard core as follows:

$$\gamma(u, v) = \begin{cases} 0 & \text{if } d(u, v) \leq R_1 \\ \gamma & \text{if } R_1 < d(u, v) \leq R_2 \\ 1 & \text{otherwise} \end{cases}$$

for u, v in a compact set $\mathcal{X} \subset \mathbb{R}^d$, $\gamma > 1$ and $R_2 > R_1 > 0$. The hard core R_1 ensures an upper bound $n(R_1, \mathcal{X})$ on the number of points that can be placed in $\mathcal{X}$ in such a way that no point is closer than R_1 to another point. Therefore, provided the intensity function $\beta(\cdot)$ is integrable, (4.1) with the Saunders–Kryscio–Funk interaction function is well-defined.

Example 4.5   The *saturation model* [63] is similar in spirit to example 4.4 in that it modifies the Strauss interaction function (4.2) to allow for

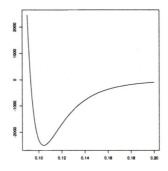

Fig. 4.4   The graph of the Lennard–Jones potential $f(r) = \alpha \left(\frac{\sigma}{d(u,v)} \right)^m - \beta \left(\frac{\sigma}{r} \right)^n$ with $\alpha = 100$, $\beta = 1$ and $\sigma = 0.2$.

clustered behaviour. Here, instead of introducing a hard core, the number of neighbour pairs is truncated. Thus, for each $x \in \mathbf{x}$, define

$$m_x(\mathbf{x}) = \sum_{\xi \in \mathbf{x} \setminus \{x\}} \mathbb{I}\{d(x,\xi) \le R\}$$

so that the sum of the m_xs is twice the number of neighbour pairs $s(\mathbf{x})$ used in the definition of the Strauss process. Writing

$$t(\mathbf{x}) = \sum_{x \in \mathbf{x}} \min[m_x(\mathbf{x}), c]$$

for some constant $c > 0$,

$$p(\mathbf{x}) = \alpha \, \beta^{n(\mathbf{x})} \, \gamma^{t(\mathbf{x})}$$

is a density for the saturation process. For $\gamma < 1$, there is repulsion between the points as for the Strauss model; for $\gamma > 1$, realisations of $p(\cdot)$ tend to be clustered. Because of the truncation in $t(\cdot)$, explosion is prevented [63] and the saturation model is well-defined for all values $\gamma \ge 0$. However, since $t(\mathbf{x})$ depends on all points of $\mathbf{x}$, the saturation model is not a pairwise interaction model.

Example 4.6   Step functions

$$\gamma(u,v) = \begin{cases} \sum_{k=1}^{K} \gamma_k \, \mathbb{I}\{R_k \le d(u,v) < R_{k+1}\} & \text{if } d(u,v) < R_{K+1} \\ 1 & \text{if } d(u,v) \ge R_{K+1} \end{cases}$$

for $u, v \in \mathcal{X}$, a compact subset of $\mathbb{R}^d$, and $R_1 = 0 < R_2 < \cdots R_K < R_{K+1}$, are often used as an approximation to more complicated interaction functions as well as in non-parametric inference [90]. A sufficient condition for integrability is that *either* all $\gamma_i \le 1$ *or* $\gamma_1 = 0$. The resulting model is Markov at range R_{K+1}.

The pairwise interaction models described above all exhibit some inhibition term (an upper bound on the second order interaction function, a truncation or a hard core) to ensure integrability. This seems to be a class trait, making pairwise interaction modelling particularly appealing for ordered point patterns.

Of course, the functional forms described here may easily be adapted to marked point processes (cf. example 2.9). For instance, the Lennard–Jones interaction function for a point at u labelled k and a point at v with label l would be [69; 156]

$$\gamma_{kl}(d(u,v)) = \exp\left[\alpha_{kl}\left(\frac{\sigma_{kl}}{d(u,v)}\right)^6 - \beta_{kl}\left(\frac{\sigma_{kl}}{d(u,v)}\right)^{12}\right]$$

for $\alpha_{kl} \ge 0$, $\beta_{kl}, \sigma_{kl} > 0$, and marks k, l in some finite set $\mathcal{K}$ of labels.

4.3   Area-interaction processes

In the previous section, the Strauss process was defined by requiring the conditional intensity to be a function of the number of neighbour pairs. When the points of $\mathcal{X}$ represent physical objects, for instance in image analysis applications [7], it is more natural to ask for $\lambda(u; \mathbf{x})$ to be defined in terms of the amount of overlap between the newly added object represented by u and those already present in $\mathbf{x}$. In the theorem below [77], the objects are balls of radius R.

Theorem 4.3   *Let $\mathcal{X} \subseteq \mathbb{R}^2$ be a bounded Borel set containing an open ball of radius $3R$. Then the area-interaction density $p(\mathbf{x}) = \alpha \, \beta^{n(\mathbf{x})} \, \gamma^{-\mu(U_\mathbf{x})}$*

for $\alpha, \beta, \gamma > 0$ *is the only function* $p : N^{\mathrm{f}} \to (0, \infty)$ *for which*

$$\frac{p(\mathbf{x} \cup \{u\})}{p(\mathbf{x})} = g\left(\mu(B(u, R) \cap U_{\mathbf{x}})\right)$$

for some left-continuous function $g : [0, \pi R^2] \to (0, \infty)$. *Here* $U_{\mathbf{x}} = \bigcup_{x \in \mathbf{x}}$ $B(x, R)$ *denotes the union of balls of radius* R *centred at the points of the configuration* $\mathbf{x}$, $n(\mathbf{x})$ *is the cardinality of* $\mathbf{x}$, *and* $\mu(\cdot)$ *is 2–dimensional Lebesgue measure.*

Note that although the characterisation theorem 4.3 is stated for the planar case only, generalisations to higher dimensions are straightforward. For the proof, we need the following lemma [77].

Lemma 4.1 *If* $g : [0, \pi R^2] \to (0, \infty)$ *is left-continuous and* $\frac{g(0)\,g(s+t)}{g(s)\,g(t)} = 1$ *for all* $s, t \in [0, \pi R^2]$ *such that* $s + t \in [0, \pi R^2]$, *then* $g(s) = g(0)\gamma^s$ *for some* $\gamma > 0$ *and all* $s \in [0, \pi R^2]$.

Proof. Without loss of generality, assume $R = 1$. Extend the function $g(\cdot)$ onto $[0, \infty)$ as follows:

$$g(k\pi + s) = \left(\frac{g(\pi)}{g(0)}\right)^k g(s), \quad s \in (0, \pi], \, k \in \mathbb{N}.$$

Then $g(\cdot)$ is left-continuous on $\mathbb{R}^+$. Moreover, for $s, t \in [0, \pi]$ with $s + t > \pi$,

$$\frac{g(0)\,g(s+t)}{g(s)\,g(t)} = \frac{g(\pi)\,g(s+t-\pi)}{g(s)\,g(t)}.$$

Now, choose $c_1, c_2 > 0$ such that $s - c_1\pi \geq 0$, $t - c_2\pi \geq 0$, and $c_1 + c_2 = 1$. Then, since $s + t - \pi = (s - c_1\pi) + (t - c_2\pi) \leq \pi$,

$$\begin{aligned}
\frac{g(0)\,g(s+t)}{g(s)\,g(t)} &= \frac{g(\pi)\,g(s - c_1\pi)\,g(t - c_2\pi)}{g(s)\,g(t)\,g(0)} \\
&= \frac{g(\pi)\,g(s - c_1\pi)\,g(t - c_2\pi)g(0)\,g(0)}{g(c_1\pi)\,g(s - c_1\pi)\,g(c_2\pi)\,g(t - c_2\pi)\,g(0)} \\
&= \frac{g(\pi)\,g(0)}{g(c_1\pi)\,g(c_2\pi)} = 1.
\end{aligned}$$

Thus, the factorisation $g(0)g(s + t) = g(s)g(t)$ holds for all $s, t \in [0, \pi]$.

For general s and t, write $s = k\pi + \tilde{s}$, $t = l\pi + \tilde{t}$ with $k, l \in \mathbb{N}$ and $\tilde{s}, \tilde{t} \in [0, \pi]$. Then

$$g(0)g(s+t) = g(0)\left(\frac{g(\pi)}{g(0)}\right)^{k+l} g(\tilde{s}+\tilde{t}) = \left(\frac{g(\pi)}{g(0)}\right)^{k+l} g(\tilde{s})g(\tilde{t}) = g(s)g(t).$$

Hence, by Hamel's theorem, $g(t) = g(0)e^{-\lambda t}$ where $\lambda = \frac{-1}{\pi}\log\frac{g(\pi)}{g(0)}$. In particular for all $s \in [0, \pi]$,

$$g(s) = g(0)\gamma^s, \quad \gamma = \left(\frac{g(\pi)}{g(0)}\right)^{1/\pi} > 0.$$

$\square$

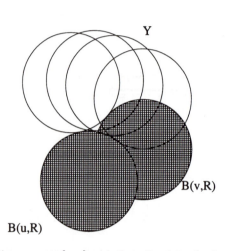

Fig. 4.5   Configuration $\mathbf{z} = \mathbf{y} \cup \{u, v\}$ such that all points of $\mathbf{y}$ have distance $2R$ to u, and balls of radius R centred at the points of $\mathbf{y}$ cover a prespecified fraction of $B(v, R)$.

We are now ready to give the proof of theorem 4.3.

Proof.   It is clear that the conditional intensity of the area-interaction model has the required form. To prove the reverse statement, take $s, t \in [0, \pi R^2]$ such that $s + t \in [0, \pi R^2]$. We will show that

$$g(s+t)\,g(0) = g(s)\,g(t). \tag{4.8}$$

Assume without loss of generality that $s \neq 0$ and $t \neq 0$, and fix $\epsilon > 0$. Since $\mathcal{X}$ contains a ball B of radius $3R$, there exist points u (the center of B) and $v \in \mathcal{X}$, and a configuration $\mathbf{y} \in N^{\mathrm{f}}$ such that

- $\mu\left(B(u,R) \cap B(v,R)\right) = s;$
- $\mu(D) = t' \in [t - \epsilon, t]$, writing $D = B(v,R) \cap U_{\mathbf{y}};$
- $\|y - u\| = 2R$ for all $y \in \mathbf{y}$

(cf. Figure 4.5). Here we use the Heine-Borel theorem to ensure that the configuration $\mathbf{y}$ is finite and $U_{\mathbf{y}}$ covers an area t of $B(v,R)$ up to the given precision ϵ without covering any area of $B(u,R)$.

Note that

$$
\begin{aligned}
g(s)\,g(t') &= \frac{p(\{u,v\} \cup \mathbf{y})}{p(\{v\} \cup \mathbf{y})}\frac{p(\{v\} \cup \mathbf{y})}{p(\mathbf{y})} \\
&= \frac{p(\{u,v\} \cup \mathbf{y})}{p(\{u\} \cup \mathbf{y})}\frac{p(\{u\} \cup \mathbf{y})}{p(\mathbf{y})} = g(s+t')\,g(0).
\end{aligned}
$$

Letting $\epsilon \downarrow 0$ and using the fact that $g(\cdot)$ is left-continuous, one obtains the desired factorisation (4.8). Consequently, by Lemma 4.1, $g(s) = g(0)\,\gamma^s$ for some $\gamma > 0$.

Next, set $\beta = g(0)\gamma^{\pi R^2} = g(\pi R^2)$. Then $g(s) = \beta\,\gamma^{s - \pi R^2}$, and hence

$$
p(\{u\}) = p(\emptyset)\,g(0) = p(\emptyset)\,\beta\,\gamma^{-\pi R^2} = p(\emptyset)\,\beta\,\gamma^{-\mu(U_{\{u\}})} = \alpha\,\beta\,\gamma^{-\mu(U_{\{u\}})},
$$

writing $\alpha = p(\emptyset)$. As an aside, remark that here it is crucial that $\mu(\cdot)$ is not restricted to $\mathcal{X}$. Finally, if $p(\mathbf{x}) = \alpha\,\beta^{n(\mathbf{x})}\,\gamma^{-\mu(U_{\mathbf{x}})}$ for all configurations consisting of (at most) n points, then

$$
\begin{aligned}
p(\mathbf{x} \cup \{u\}) &= g\left(\mu\left(B(u,R) \cap U_{\mathbf{x}}\right)\right) f(\mathbf{x}) \\
&= g(0)\gamma^{\mu(B(u,R) \cap U_{\mathbf{x}})}\alpha\beta^n\gamma^{-\mu(U_{\mathbf{x}})} \\
&= \alpha\beta^{n+1}\gamma^{-\pi R^2 + \mu(B(u,R)\cap U_{\mathbf{x}}) - \mu(U_{\mathbf{x}})} \\
&= \alpha\beta^{n+1}\gamma^{-\mu(U_{\mathbf{x} \cup \{u\}})}.
\end{aligned}
$$

$\square$

The parameter γ in theorem 4.3 can be interpreted in the following way. If $\gamma > 1$, configurations $\mathbf{x}$ for which the area of $U_{\mathbf{x}}$ is small are favoured, so the points of $\mathbf{x}$ tend to cluster together. If $\gamma < 1$, on the other hand, regular configurations whose associated objects cover a large area are more likely [8]. The special case $\gamma = 1$ corresponds to a Poisson process. In contrast to the models discussed in Section 4.2, the area–interaction model dependencies are not restricted to pairs of overlapping objects. Indeed, by the inclusion–exclusion formula, the area of $U_{\mathbf{x}}$ can be written as

$$\sum_{k=1}^{n} \mu(B(x_k, R)) - \sum_{k<l} \mu(B(x_k, R) \cap B(x_l, R)) \cdots (-1)^{n+1} \mu(\cap_{i=1}^{n} B(x_i, R))$$

if $\mathbf{x} = \{x_1, \ldots, x_n\}$. Hence, the area-interaction process is Markov at range $2R$ with clique interaction function given by [8]

$$\phi(\emptyset) = p(\emptyset)$$
$$\phi(\{u\}) = \beta \gamma^{-\pi R^2}$$
$$\phi(\{x_1, \ldots, x_n\}) = \gamma^{(-1)^n \mu(\cap_{k=1}^{n} B(x_k, R)))} \quad \text{for } n \geq 2$$

(cf. theorem 2.2).

In example 2.11, it was shown that an attractive area-interaction model can be obtained as the marginal distribution of a component in the penetrable spheres mixture model (cf. example 2.8). A similar interpretation in terms of Poisson processes holds for the inhibitory case [8; 221].

Example 4.7 Let X and Y be independent Poisson processes on a compact subset $\mathcal{X} \subseteq \mathbb{R}^d$ with rates $\beta > 0$ and $|\log \gamma|$, $\gamma > 0$, respectively. Assume the volume $\mu(\mathcal{X})$ is positive and finite.

In the inhibitory case $\gamma < 1$,

$$\mathbf{P}(Y \subseteq U_X | X) = \mathbf{P}(Y \cap (\mathcal{X} \setminus U_X) = \emptyset) = \gamma^{\mu(\mathcal{X})} \gamma^{-\mu(U_X)}$$

hence $\mathbf{P}(Y \subseteq U_X) = \mathbf{E}_{\beta} \left[\gamma^{\mu(\mathcal{X})} \gamma^{-\mu(U_X)} \right]$, the expectation of $\gamma^{\mu(\mathcal{X})} \gamma^{-\mu(U_X)}$ under the distribution of X. Thus, the conditional distribution of X given $\{Y \subseteq U_X\}$ has a density proportional to $\gamma^{-\mu(U_{\mathbf{x}})}$ with respect to a Poisson process of rate β, or proportional to $\beta^{n(\mathbf{x})} \gamma^{-\mu(U_{\mathbf{x}})}$ if the dominating measure is a unit rate Poisson process. Hence, conditional on $\{Y \subseteq U_X\}$, X is an area-interaction process.

For $\gamma > 1$, $\mathbf{P}(Y \cap U_X = \emptyset | X) = e^{-\mu(U_X) \log \gamma} = \gamma^{-\mu(U_X)}$. Hence conditional on $\{Y \cap U_X = \emptyset\}$, X is an area-interaction model in accordance with the conclusion of example 2.11.

The penetrable spheres representation is particularly convenient from a sampling point of view. Of course, since the area-interaction density is locally stable, algorithm 3.4 may be used to obtain realisations. However, an auxiliary variables algorithm exploiting the connection with the penetrable spheres mixture model is a good alternative.

More precisely, suppose $\beta > 0$, $\gamma > 1$, and let $(\mathbf{x}, \mathbf{y})$ be some initial state. Then, if $Z_{1,x}$ and $Z_{1,y}$ are two independent Poisson processes on $\mathcal{X}$ with rates β and $\log \gamma$ respectively, a single iteration of the Gibbs sampler results in a new state (X_1, Y_1) where $X_1 = Z_{1,x} \setminus U_{\mathbf{y}}$ and $Y_1 = Z_{1,y} \setminus U_{X_1}$. Clearly, if $\mathbf{y}' \subseteq \mathbf{y}$,

$$X_1' = Z_{1,x} \setminus U_{\mathbf{y}'} \supseteq Z_{1,x} \setminus U_{\mathbf{y}} = X_1$$

and a fortiori $U_{X_1'} \supseteq U_{X_1}$ so that

$$Y_1' = Z_{1,y} \setminus U_{X_1'} \subseteq Z_{1,y} \setminus U_{X_1} = Y_1.$$

Thus, the Gibbs sampler preserves the partial order

$$(\mathbf{x}, \mathbf{y}) \preceq (\mathbf{x}', \mathbf{y}') \Leftrightarrow \mathbf{x} \subseteq \mathbf{x}' \text{ and } \mathbf{y} \supseteq \mathbf{y}'.$$

Although the partial order $\preceq$ does not have a maximum in N^{f}, the Gibbs sampler only uses the union sets U_X, U_Y, and we can define *quasimaximal* states as those bivariate patterns $(\mathbf{x}, \mathbf{y})$ for which $\mathbf{y} = \emptyset$ and $U_{\mathbf{x}} \supseteq \mathcal{X}$. Similarly, $(\mathbf{x}, \mathbf{y})$ is called *quasiminimal* if $\mathbf{x} = \emptyset$ and $U_{\mathbf{y}} \supseteq \mathcal{X}$. Coupling a Gibbs sampler starting in any quasimaximum to one starting in some quasiminimum yields an exact simulation algorithm [77].

Algorithm 4.1 Suppose $\beta > 0$ and $\gamma > 1$ are given. Initialise $T = 1$, and let $(\mathbf{x}, \mathbf{y})$ be a quasiminimal, $(\mathbf{x}', \mathbf{y}')$ a quasimaximal state. For $n = -1, -2, \ldots$, let $Z_{n,x}$ and $Z_{n,y}$ be independent Poisson processes on $\mathcal{X}$ with respective intensities β and $\log \gamma$. Repeat

- initialise $(X_{-T}(-T), Y_{-T}(-T)) = (\mathbf{x}, \mathbf{y})$, $(X_{-T}'(-T), Y_{-T}'(-T)) = (\mathbf{x}', \mathbf{y}')$;
- for $n = -T$ to -1, set $X_{-T}(n+1) = Z_{n,x} \setminus U_{Y_{-T}(n)}$ and $Y_{-T}(n+1) = Z_{n,y} \setminus U_{X_{-T}(n)}$, $X_{-T}'(n+1) = Z_{n,x} \setminus U_{Y_{-T}'(n)}$ and $Y_{-T}'(n+1) = Z_{n,y} \setminus U_{X_{-T}'(n)}$;
- if $(X_{-T}(0), Y_{-T}(0)) = (X_{-T}'(0), Y_{-T}'(0))$, return the common value; otherwise set $T := 2T$;

until the upper and lower processes have coalesced.

It can be shown that algorithm 4.1 terminates almost surely in finite time, yielding a sample from the penetrable spheres mixture model [77]. The first component is a realisation of the area-interaction process with parameters β and γ.

Theorem 4.4   *Let* $\mathcal{X} \subseteq \mathbb{R}^d$ *be a compact set with finite volume* $\mu(\mathcal{X}) > 0$, $\beta > 0$ *and* $\gamma > 1$. *Then algorithm 4.1 terminates almost surely, and returns a sample from the penetrable spheres mixture model defined by its density*

$$p(\mathbf{x}, \mathbf{y}) = \alpha\, \beta^{n(\mathbf{x})}\, (\log \gamma)^{n(\mathbf{y})}\; \mathrm{I\!I}\{d(\mathbf{x}, \mathbf{y}) > R\}$$

with respect to the product of two unit rate Poisson process on $\mathcal{X}$.

Proof.   Note that if $Z_{n,x} = \emptyset$ for some $n \in \{-1, -2, \dots\}$, the upper and lower processes merge at time $n + 1$. Since this happens with probability 1, algorithm 4.1 terminates almost surely.

To show that the output has the required distribution, let I be the stopping time defined as the smallest n for which $Z_{n,x} = \emptyset$ and consider the Markov chains (X''_{-T}, Y''_{-T}) governed by the same transitions as (X_{-T}, Y_{-T}) and (X'_{-T}, Y'_{-T}) but initialised by a sample from the equilibrium distribution, independently of all other random variables. Then for $-T \leq I$, $X_{-T}(I+1) = X'_{-T}(I+1) = X''_{-T}(I+1) = \emptyset$, $Y_{-T}(I+1) = Y'_{-T}(I+1) = Y''_{-T}(I+1) = Z_{I,y}$. Hence (X_{-T}, Y_{-T}), (X'_{-T}, Y'_{-T}) and (X''_{-T}, Y''_{-T}) yield the same output at time 0, so $(X(0), Y(0)) = \lim_{T \to \infty}(X''_{-T}(0), Y''_{-T}(0))$ exists almost surely. Since $X''_{-T}(\cdot)$ is in equilibrium, the distribution of $(X(0), Y(0))$ is that of the penetrable spheres mixture model with parameters β and $\log \gamma$. Finally, $X(0)$ is an area-interaction process with parameters β and γ (cf. example 2.11).   □

Figure 2.2 was obtained by means of algorithm 4.1. Coalescence was reached in 8 steps. A modification to take into account the subtle bias caused by the user breaking off long runs [55] can be found in [216].

4.4 Shot noise and quermass-interaction processes

The area interaction model (section 4.3) has been generalised in various directions. First of all, instead of placing a ball of fixed radius around each point, a compact set $Z(x)$ may be assigned to each point $x \in \mathbf{x}$ before forming the union $U_{\mathbf{x}} = \bigcup_{x \in \mathbf{x}} Z(x)$ (assuming that the mapping $u \in \mathcal{X} \mapsto Z(u)$ is sufficiently smooth to ensure the resulting density is measurable [8]). This approach is particularly convenient when the interaction structure is not homogeneous over the region of interest. Indeed, points are neighbours whenever their associated sets overlap. Similarly, the Lebesgue measure $\mu(\cdot)$ may be replaced by any finite, Borel regular measure $\nu(\cdot)$ on $\mathcal{X}$, and the intensity parameter β may be replaced by a Borel measurable intensity function reflecting spatial variation. Doing so, one obtains the *general area-interaction process* [8] with density

$$p(\mathbf{x}) = \alpha \prod_{x \in \mathbf{x}} \beta(x) \, \gamma^{-\nu(U(\mathbf{x}))}$$

with respect to a unit rate Poisson process on $\mathcal{X}$. The process is Markov with respect to the overlapping objects relation [6]

$$u \sim v \Leftrightarrow Z(u) \cap Z(v) \neq \emptyset. \tag{4.9}$$

Of course, the sets $Z(\cdot)$ may be chosen randomly, yielding a marked area-interaction process which is Markov with respect to the mark dependent neighbourhood relation defined by (4.9). For $\gamma \in (0,1)$, configurations for which the ν-mass is large are favoured, resulting in more ordered patterns; for $\gamma = 1$, $p(\mathbf{x})$ defines a Poisson process, while for $\gamma > 1$ realisations tend to be more clustered than under the dominating Poisson process.

Taking a geometric point of view, the basic measurements of a planar object are its area, its perimeter and Euler characteristic. Thus, it is natural to define *quermass-interaction processes* [104; 136] by a density of the form

$$p(\mathbf{x}) = \alpha \, \beta^{n(\mathbf{x})} \exp \left[\gamma_1 \, \mu(U_{\mathbf{x}}) + \gamma_2 \, \eta(U_{\mathbf{x}}) + \gamma_3 \chi(U_{\mathbf{x}}) \right] \tag{4.10}$$

where $\eta(\cdot)$ denotes the perimeter length, $\chi(\cdot)$ is the Euler characteristic, and $\gamma_1, \gamma_2, \gamma_3$ are real valued parameters. As before $U_{\mathbf{x}} = \cup_{x \in \mathbf{x}} B(x, R)$. If $\gamma_2 = \gamma_3 = 0$, (4.10) defines an area-interaction process.

Theorem 4.5 *Let $\mathcal{X}$ be a compact subset of $\mathbb{R}^2$ with positive, finite area*

$\mu(\mathcal{X})$. *Then the quermass-interaction process defined by its density (4.10) with respect to a unit rate Poisson process on $\mathcal{X}$ is well-defined. It is Ripley–Kelly Markov at range 2R.*

Quermass interaction processes cannot be generalised to arbitrary dimensions [104]. Indeed, in $\mathbb{R}^d$, $d \geq 4$, for $\gamma_3 < 0$ [149] it is possible to construct configurations of $2n$ balls with Euler characteristic $-n^2$, so that we cannot bias towards 'holiness'. The spatial case $d = 3$ seems to be an open question.

Proof. Since $\mu(U_\mathbf{x})$ is uniformly bounded, and $0 \leq \eta(U_\mathbf{x}) \leq 2\pi\, n(\mathbf{x})\, R$, the function $\beta^{n(\mathbf{x})} \exp\left[\gamma_1\, \mu(U(\mathbf{x})) + \gamma_2\, \eta(U_\mathbf{x})\right]$ is integrable with respect to the distribution of a unit rate Poisson process. Thus, it suffices to show that the term $\exp\left[\gamma_3 \chi(U_\mathbf{x})\right]$ is integrable as well.

Now, since the planar Euler characteristic is the difference between the number of components and the number of holes, $\chi(U_\mathbf{x})$ does not exceed the cardinality of $\mathbf{x}$. Hence, for $\gamma_3 \geq 0$,

$$\exp\left[\gamma_3\, \eta(U_\mathbf{x})\right] \leq (e^{\gamma_3})^{n(\mathbf{x})}.$$

The right hand side is proportional to the density of a Poisson process of rate e^{γ_3}, from which we conclude that the quermass-interaction density is integrable.

To show integrability for $\gamma_3 < 0$, note that each component of the complement $U_\mathbf{x}^c$ must contain at least one node (possibly the one at infinity) of the Voronoi tessellation of $\mathbf{x}$, for otherwise the boundary of this component would have to be made out of the boundaries of at most two discs. Hence the number of holes in $U_\mathbf{x}$ is bounded by the number of nodes of the tessellation, or equivalently, by the number of triangles in the Delaunay triangulation formed by connecting Voronoi cells sharing a common edge [142] as in Figure 4.6. The latter number is bounded above by $2(n(\mathbf{x}) - 1)$ [142, p. 24], hence integrablility is ensured.

Finally, turn to the Markov property. By the inclusion-exclusion formula, for any configuration $\mathbf{x}$ and $u \in \mathcal{X} \setminus \mathbf{x}$, the conditional intensity at u given $\mathbf{x}$ equals

$$\lambda(u, \mathbf{x}) = \beta \exp\left[\gamma_1 \mu(B(u, R)) - \gamma_1 \mu(B(u, R) \cap U_\mathbf{x}) + \gamma_2 \eta(B(u, R))\right]$$

$$\times \exp\left[-\gamma_2 \eta(B(u,R) \cap U_{\mathbf{x}}) + \gamma_3 \chi(B(u,R)) - \gamma_3 \chi(B(u,R) \cap U_{\mathbf{x}})\right].$$

In the above formula, one may replace $U_{\mathbf{x}}$ by $U_{\mathbf{x}(u)}$ where $\mathbf{x}(u)$ is the configuration consisting of all points $x \in \mathbf{x}$ such that $B(u,R) \cap B(x,R) \neq \emptyset$. It follows that $\lambda(u,\mathbf{x})$ depends only on u and its $2R$-neighbours, and the proof is complete.   □

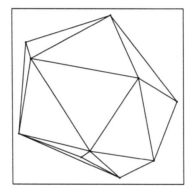

Fig. 4.6   Delaunay triangulation based on 10 Voronoi cells.

As for area-interaction, (4.10) can easily be adapted to a marked quermass-interaction process [104].

A third approach is based on the observation that the area-interaction model only considers the total area covered by $U_{\mathbf{x}}$. However, it may well be of interest to take the multiplicity, that is the number of balls covering a point of $\mathcal{X}$, into account [124]. Thus, define the *coverage function* of $\mathbf{x}$ by

$$c_{\mathbf{x}}(u) = \sum_{x \in \mathbf{x}} \mathbb{I}\{u \in B(x,R)\},$$

that is, $c_{\mathbf{x}}(u)$ counts the number of balls covering the point u. The function

$c_{\mathbf{x}}(u)$ is symmetric in the points of $\mathbf{x}$ and Borel measurable. Then, a *shot noise weighted point process with potential function* $f(\cdot)$ has density

$$p(\mathbf{x}) = \alpha\,\beta^{n(\mathbf{x})}\,\gamma^{-\int_{\mathcal{X}} f(c_{\mathbf{x}}(u))\,d\mu(u)}. \tag{4.11}$$

Here $\beta, \gamma > 0$ are model parameters, and $f : \mathbb{N}_0 \to \mathbb{R}$ is a Borel measurable function such that $f(0) = 0$ and

$$|f((c_{\mathbf{x}}(u))| \le C n(\mathbf{x})$$

for some positive constant C. The latter condition ensures that (4.11) is integrable [124]. Note that if $f(n) = \mathbb{I}\{n \ge 1\}$, (4.11) is the density of an area-interaction process. The shot noise weighted point process is Ripley-Kelly Markov at range $2R$. If balls are replaced by $Z(u)$, $u \in \mathcal{X}$, the process is Markov with respect to the overlapping objects relation (4.9).

Finally, note that (4.11) is overparameterised, since taking $f'(\cdot) = cf(\cdot)$ for some constant c is equivalent to changing γ to γ^c. If $f(\cdot)$ is integrable in absolute value, this ambiguity can be overcome by requiring that $f(\cdot)$ integrates to unity.

Example 4.8 Let $f(n) = \mathbb{I}\{n = 1\}$. Then (4.11) reads

$$p(\mathbf{x}) = \alpha\,\beta^{n(\mathbf{x})}\,\gamma^{-\mu(\{u \in \mathcal{X}: c_{\mathbf{x}}(u)=1\})}.$$

For $\gamma > 1$ the model tends to have smallish 1–covered regions, so that aggregated patterns are likely. For $\gamma < 1$, largish 1–covered regions or more regular patterns are favoured. As the quermass-interaction process (4.10), $p(\mathbf{x})$ exhibits interactions of any order [124].

Example 4.9 Set $f(n) = \mathbb{I}\{n = 2\}$. Then $\int_{\mathcal{X}} f(c_{\mathbf{x}}(u))\,d\mu(u)$ is the volume of the set $U_2(\mathbf{x})$ of points in $\mathcal{X}$ covered by exactly two balls $B(x, R)$, $x \in \mathbf{x}$, and, therefore,

$$p(\mathbf{x}) = \alpha\,\beta^{n(\mathbf{x})}\,\gamma^{-\mu(U_2(\mathbf{x}))}.$$

For $\gamma > 1$ there tend to be many high order overlaps or no overlaps at all, while, for $\gamma < 1$, objects tend to come in pairs [124].

4.5 Morphologically smoothed area-interaction processes

In the previous sections, we saw that smoothness constraints have to be imposed on the marks in a marked quermass- or area-interaction process. One way of enforcing such smoothness is by means of morphological operators [88; 130; 195].

Let A and B be subsets of $\mathbb{R}^d$, and write $B_h = h + B$ for the translation of B over a vector $h \in \mathbb{R}^d$. Then the *morphological opening* of A by the *structuring element* B is defined as [130]

$$A \circ B = \bigcup \{ B_h : h \in \mathbb{R}^d, B_h \subseteq A \};$$ (4.12)

the *morphological closing* of A by B is the dual operation

$$A \bullet B = (A^c \circ B)^c.$$ (4.13)

Thus, the opening operator removes those parts of A that are 'smaller' than B, while closing has the effect of filling small gaps.

Example 4.10 Let $\mathcal{X}$ be a compact subset of $\mathbb{R}^d$ and write $\mathcal{K}$ for the family of compact sets. Define a marked point process on $\mathcal{X} \times \mathcal{K}$ by its density

$$p(\mathbf{y}) = \alpha \gamma^{-\mu(\mathcal{X} \cap (U_{\mathbf{y}} \circ B))}$$ (4.14)

with respect to the distribution of a unit rate Poisson process on $\mathcal{X}$ marked independently according to a probability distribution $m(\cdot)$ on $\mathcal{K}$. Here, α is the normalising constant, $\gamma > 0$ is the interaction parameters, and $B \neq \emptyset$ is a convex, compact subset of $\mathbb{R}^d$. As usual, $U_{\mathbf{y}} = \bigcup_{(x,K) \in \mathbf{y}} K_x$ denotes the union of particles represented by $\mathbf{y}$. The process defined by (4.14) is called an *opening-smoothed area-interaction process*. The area-interaction point process is the special case where the mark distribution is degenerate, and $B = \{0\}$.

Similarly,

$$p(\mathbf{y}) = \alpha \gamma^{-\mu(\mathcal{X} \cap (U_{\mathbf{y}} \bullet B))}$$ (4.15)

defines a *closing-smoothed area-interaction process* [120].

As an aside, the mark space $\mathcal{K}$ with the myope topology is locally compact and second countable [130], hence can be equipped with a complete metric [17]. Clearly, $p(\mathbf{y})$ is symmetric in the members of $\mathbf{y}$, and uniformly bounded. It can be shown that (4.14)–(4.15) are measurable, but to do so we would need more details about the topological structure of $\mathcal{K}$. The interested reader is referred to chapter 1 of [130].

In (4.14)–(4.15), Lebesgue measure may be replaced by e.g. the Euler characteristic. Note that due to the smoothing effect, (4.14)–(4.15) may be integrable when their non-smoothed counterparts are not.

Let $X_1, \cdots, X_m$ denote the connected components of $U_\mathbf{y}$. Then $U_\mathbf{y} \circ B = \cup \{B_h \subseteq U_\mathbf{y}\} = \cup_{i=1}^m X_i \circ B$, since the convexity of B implies that B_h must be contained entirely in a single component. Hence (4.14) can be written as

$$\alpha \prod_{i=1}^m \Phi(X_i)$$

where $\Phi(K) = \gamma^{-\mu(\mathcal{X} \cap (K \circ B))}$, which implies that the marked point process Y defined by (4.14) is connected component Markov.   By duality, the closing $U_\mathbf{y} \bullet B$ distributes over the connected components X_i^c, $i = 1, \ldots, l$, of $\mathcal{X} \setminus U_\mathbf{y}$, so $\mu(\mathcal{X} \cap (U_\mathbf{y} \bullet B)) = \mu(\mathcal{X} \setminus (\cup_{i=1}^l X_i^c \circ B))$, hence (4.15) is equivalent to

$$\alpha \prod_{i=1}^l \Psi(X_i^c)$$

where $\Psi(K) = \gamma^{\mu(\mathcal{X} \cap (K \circ B))}$.   It is interesting to note that the above factorisations remain valid if only the union set $U_\mathbf{y}$ were observed, (cf. example 1.12) a remark leading naturally to the notion of a Markov random set [118; 120].

Example 4.11   Let $\mathcal{X}$ be a compact subset of $\mathbb{R}^d$ and write $\mathcal{K}$ for the family of compact sets. Then a *morphologically smooth marked point process* has density

$$p(\mathbf{y}) = \alpha \gamma^{-\mu((\mathcal{X} \cap U_\mathbf{y}) \setminus (U_\mathbf{y} \circ B))} \tag{4.16}$$

with respect to the distribution of a unit rate Poisson process on $\mathcal{X}$ marked independently according to a probability distribution $m(\cdot)$ on $\mathcal{K}$, where B is a convex compact set and $\gamma > 0$ the interaction parameter. For $\gamma > 1$, the most likely realisations are those for which $U_\mathbf{y}$ is open with respect to the structuring element B, i.e. $U_\mathbf{y} = U_\mathbf{y} \circ B$. Thus, configurations $\mathbf{y}$ for which $U_\mathbf{y}$ is similar in shape to an ensemble of Bs are favoured over those with thin or elongated pieces, sharp edges or small isolated clutter [33; 70; 120; 201; 202]. By duality

$$p(\mathbf{y}) = \alpha \gamma^{-\mu(\mathcal{X} \cap (U_\mathbf{y} \bullet B) \setminus U_\mathbf{y})}, \tag{4.17}$$

which for $\gamma > 1$ favours sets that are approximately closed with respect to B, and discourages small holes or rough edges. Both (4.16) and (4.17) are well-defined for $\gamma \leq 1$ too, in which case morphological roughness is encouraged.

As the opening-smoothed area-interaction process, the opening-smooth marked point process is connected component Markov, but a Hammersley–Clifford factorisation in the sense of Ripley and Kelly holds as well [120; 202]. In order to state the result, the following concepts from mathematical morphology are needed. For $A, B \subseteq \mathcal{X}$, the *Minkowski addition* $A \oplus B$ is the set $\{a + b : a \in A, b \in B\}$; furthermore, $\check{B} = \{-b : b \in B\}$ is the *reflection* of B in the origin.

Theorem 4.6 *Morphologically smooth marked point processes are Markov with respect to the mark-dependent relation*

$$(u, K) \sim (v, L) \Leftrightarrow (K_u \oplus (B \oplus \check{B})) \cap (L_v \oplus (B \oplus \check{B})) \neq \emptyset$$

on $\mathcal{X} \times \mathcal{K}$, where $\mathcal{X}$ is a compact subset of $\mathbb{R}^d$ and $\mathcal{K}$ the family of compact sets.

Proof. Let $\mu(\cdot)$ be Lebesgue measure restricted to $\mathcal{X}$, and consider the opening-smooth marked point process defined by (4.16). Then the Papangelou conditional intensity for adding (u, K) to a configuration $\mathbf{y}$ equals

$$\lambda((u, K); \mathbf{y}) = \gamma^{-\mu(U_\mathbf{y} \cup K_u) \setminus ((U_\mathbf{y} \cup K_u) \circ B)) + \mu(U_\mathbf{y} \setminus (U_\mathbf{y} \circ B))}.$$

Since for $x \notin K_u \oplus (B \oplus \check{B})$,

$$x \in U_\mathbf{y} \circ B \Leftrightarrow x \in (U_\mathbf{y} \cup K_u) \circ B,$$

only the points in $K_u \oplus (B \oplus \check{B})$ contribute to the μ-mass in the exponent of $\lambda((u, K); \mathbf{y})$. By the local knowledge principle [195] of mathematical morphology, in order to evaluate $U_{\mathbf{y}}$, $U_{\mathbf{y}} \circ B$ and $(U_{\mathbf{y}} \cup K_u) \circ B$ on $K_u \oplus (B \oplus \check{B})$, it is sufficient to know the intersection of $U_{\mathbf{y}}$ with $(K_u \oplus (B \oplus \check{B})) \oplus (B \oplus \check{B})$ and the proof is complete. The equivalent statement for the closing can be derived analogously.   □

4.6   Hierarchical and transformed processes

In the previous sections, much attention was paid to spatial interaction. However, it is important to realise that spatial variation may also influence the appearance of a point pattern.

Let us, for concreteness' sake, consider a Markov point process on some space $\mathcal{X}$ with density $p(\cdot)$ with respect to a unit rate Poisson process. Then by the Hammersley–Clifford theorem, for $\mathbf{x} \in N^{\mathrm{f}}$, $p(\mathbf{x})$ can be written as

$$p(\mathbf{x}) = p(\emptyset) \prod_{x \in \mathbf{x}} \phi(\{x\}) \prod_{\text{cliques } \mathbf{y} \subseteq \mathbf{x}: n(\mathbf{y}) \geq 2} \phi(\mathbf{y}).$$

The first term $p(\emptyset)$ is the normalising constant needed to make sure that $p(\cdot)$ integrates to unity, the product over cliques with two or more points describes the inter-point interaction, and the first order terms $\phi(\{x\})$ reflect the spatial variation. Note that if the dominating Poisson process were replaced by a Poisson process with intensity function $\phi(\{x\})$, $x \in \mathcal{X}$, the factorisation would read

$$\tilde{p}(\mathbf{x}) = \tilde{p}(\emptyset) \prod_{\text{cliques } \mathbf{y} \subseteq \mathbf{x}: n(\mathbf{y}) \geq 2} \phi(\mathbf{y}).$$

Thus, inhomogeneity can be modelled by means of the first order interaction function or the intensity measure of the dominating Poisson process. For example, a realisation of an inhomogeneous Poisson process with exponential decay in one of the coordinates can be seen in Figure 1.7. More generally, [157] suggest a log polynomial intensity function in the context of a biological data set, and statistical physicists [191] express the force of an external field on a collection of interacting particles by means of the singleton interaction function.

Recently, Jensen and Nielsen [94] proposed to consider transformations of a homogeneous point process. Thus, let $h : \mathcal{X} \to \mathcal{Y}$ be an invertible, differentiable mapping between complete, separable metric spaces, and consider a Markov point process X (with respect to some neighbourhood relation $\sim$) on $\mathcal{X}$ specified by a density $p_X(\cdot)$ with respect to the unit rate Poisson process on $\mathcal{X}$. Then the image of X under h has density

$$p_Y(\mathbf{y}) = \alpha p_X(h^{-1}(\mathbf{y})) \prod_{y \in \mathbf{y}} J_h^{-1}(y) \tag{4.18}$$

with respect to the unit rate Poisson process on $\mathcal{Y}$. Here $J_h(\cdot)$ denotes the Jacobean of $h(\cdot)$. The mapping $h(\cdot)$ induces a neighbourhood relation $\approx$ on $\mathcal{Y}$ by

$$u \approx v \Leftrightarrow h^{-1}(u) \sim h^{-1}(v).$$

It is easily seen that Y is Markov with respect to $\approx$. Moreover, its clique interaction functions are readily expressed in terms of those of X as follows. Write $\phi_X(\cdot)$ for the interaction functions of the Markov point process X. Then the interaction functions $\phi_Y(\cdot)$ of Y are

$$\begin{aligned} \phi_Y(\emptyset) &= \alpha \phi_X(\emptyset); \\ \phi_Y(\{u\}) &= \phi_X(h^{-1}(u)) J_h^{-1}(u); \\ \phi_Y(\mathbf{y}) &= \phi_X(h^{-1}(\mathbf{y})) \end{aligned}$$

for configurations $\mathbf{y}$ containing at least two points. Thus, the interactions of second and higher order are inherited from X, while the first order term is determined by the transformation via its Jacobean and may well be non-constant, even if the corresponding interaction function of X is.

It is important to note that in the induced relation interaction depends on the transformation: it is harder for points in regions where $h(\cdot)$ is 'contracting' to be neighbours than for other points. This may be quite natural in biological applications, but less so in physics. The reader should compare the situation to general area-interaction models (cf. sections 4.3–4.4), where the neighbourhood relation depended on $Z(\cdot)$.

To conclude this section, let us say a few words about hierarchical models. These occur for instance in forestry, where a region is populated by trees and smaller ferns. Now the trees may not care about the ferns, whereas the ferns depend on the trees for shelter. In such situations, one could proceed by first modelling the trees, and, conditionally on the pattern of trees, specify a probability distribution for the ferns including an intensity function that specifies the dependence on the tree configuration [197].

4.7 Cluster processes

In many biological applications, a natural model for aggregated data is a *spatial cluster process* [40; 41; 46; 208]. Here, resembling the dynamics of evolution, each object x in an (unobserved) parent process X gives rise to a finite process Z_x of daughters. The next generation

$$Y = \bigcup_{x \in X} Z_x$$

forms the cluster process. Usually, the offspring processes Z_x are assumed to be independent.

Example 4.12 Let X be a unit rate Poisson process of parents on a compact subset $\mathcal{X} \subseteq \mathbb{R}^d$. Furthermore, suppose a parent at x gives birth to a Poisson number of daughters with mean ω, positioned i.i.d. according to a probability density $h(\cdot - x)$ supported on a ball $B(x, R)$ of radius R centred at the parent. Thus, the daughter pattern is a subset of $\mathcal{X}_{\oplus R} = \{x \in \mathbb{R}^d : d(x, \mathcal{X}) \le R\}$, the R-envelope of $\mathcal{X}$. Furthermore, a density $p(\mathbf{y})$ for daughter configuration $\mathbf{y} = \{y_1, \dots, y_m\}$ is given by

$$\sum_{n=1}^{\infty} \left\{ \frac{e^{\mu(\mathcal{X}_{\oplus R} \backslash \mathcal{X})} \omega^m e^{-\omega n}}{n!} \int_{\mathcal{X}} \cdots \int_{\mathcal{X}} \prod_{j=1}^{m} \left(\sum_{i=1}^{n} h(y_j - x_i) \right) dx_1 \dots dx_n \right\}$$

with respect to the distribution of a unit rate Poisson process on $\mathcal{X}_{\oplus R}$, which can be factorised as

$$p(\mathbf{y}) \;=\; e^{\mu(\mathcal{X}_{\oplus R})} \omega^m e^{-m\omega} e^{-\beta} \left[\sum_{C_1, \dots, C_k} e^{\omega(m-k)} J(\mathbf{y}_{C_1}) \dots J(\mathbf{y}_{C_k}) \right],$$

where $\beta = (1 - e^{-\omega}) \mu(\mathcal{X})$, the sum is over all *unordered* partitions of $\mathbf{y}$ into disjoint *non-empty* subconfigurations, and

$$J(\mathbf{y}_C) = \int_{\mathcal{X}} \prod_{y_j \in \mathbf{y}_C} h(y_j - x) \, dx.$$

Note that $J(\mathbf{y}_C) = 0$ unless $(\mathbf{y}_C)_{\oplus R}$ is connected. For instance, for the Matérn cluster process [129] of example 1.11 in which the daughters are uniformly distributed in a ball around their parent, $J(\mathbf{y}_C)$ is proportional to the volume occupied within the set $\mathcal{X}$ by the intersection of balls of radius R centred at the points of $\mathbf{y}_C$. Thus, the Matérn process is nearest-neighbour Markov with respect to the connected component relation at distance $2R$.

More generally, any cluster process with Poisson parents and uniformly bounded clusters is connected component Markov [10].

Theorem 4.7 *Let $(\mathcal{X}, d)$ be a complete, separable metric space, $\nu(\cdot)$ a finite, non-atomic Borel measure, and $\pi_\nu(\cdot)$ the distribution of a Poisson process of parents on $\mathcal{X}$ with intensity measure $\nu(\cdot)$. Suppose the clusters are uniformly bounded in the sense that $Z_x \subseteq B(x, R) = \{\xi \in \mathcal{X} : d(x, \xi) \leq R\}$ almost surely for some $R > 0$ and any $x \in \mathcal{X}$. Furthermore, assume that the daughter processes Z_x are absolutely continuous with respect to a unit rate Poisson process on $\mathcal{X}$, with a density $p_x(\mathbf{y})$ that is jointly Borel measurable as a mapping from $\mathcal{X} \times N^f$ onto $\mathbb{R}^+$ and hereditary excluding the empty set. Then $Y = \bigcup_{x \in X} Z_x$ is a connected component Markov point process at range $2R$ (but not in general Markov in the Ripley–Kelly sense!).*

Proof. For $x \in \mathcal{X}$ set $q_x(\cdot) = e^{-\nu(\mathcal{X})} p_x(\cdot)$. Then the total offspring density is given by

$$p(\mathbf{y}) =$$
$$\sum_{n=1}^{\infty} \frac{e^{-\nu(\mathcal{X})}}{n!} \int_{\mathcal{X}} \cdots \int_{\mathcal{X}} e^{\nu(\mathcal{X})} \sum_{C_1,\dots,C_n} \prod_{i=1}^{n} q_{x_i}(\mathbf{y}_{C_i})\, d\nu(x_1) \cdots d\nu(x_n)$$
$$= \sum_{n=1}^{\infty} \frac{1}{n!} \sum_{C_1,\dots,C_n} \prod_{i=1}^{n} \int_{\mathcal{X}} q_x(\mathbf{y}_{C_i})\, d\nu(x) \tag{4.19}$$

for $\mathbf{y} \neq \emptyset$. Here the inner sum ranges over all *ordered* partitions of $\mathbf{y}$ into n disjoint, possibly empty, sets. For $\mathbf{y} = \emptyset$,

$$p(\emptyset) = 1 + \sum_{n=1}^{\infty} \frac{1}{n!} \left[\int_{\mathcal{X}} q_x(\emptyset)\, d\nu(x) \right]^n = e^{\nu(\mathcal{X}) - \beta}$$

where $\beta = \int_{\mathcal{X}} (1 - q_x(\emptyset))\, d\nu(x)$, the mean number of non-empty clusters.

Now $q_x(\mathbf{z}) = 0$ whenever $\mathbf{z} \not\subseteq B(x, R)$; hence if $q_\xi(\mathbf{z}) \neq 0$ then all pairs of points in $\mathbf{z}$ are $2R$-close, i.e. $\mathbf{z}$ is a clique with respect to the finite range relation with distance $2R$. Hence the integral in (4.19) is nonzero only when the partition consists of $2R$-cliques.

For $\mathbf{y} \neq \emptyset$, write $\mathbf{y}_{D_1}, \ldots, \mathbf{y}_{D_K}$ for the connected components of $\mathbf{y}$ at range $2R$. Then the integral in (4.19) is nonzero only when the partition is a refinement of $D_1, \ldots, D_K$. Let $C_1, \ldots, C_k$ be an unordered partition refining $D_1, \ldots, D_K$ and consisting of non-empty sets. This contributes a term

$$\alpha \prod_{i=1}^{k} \int_{\mathcal{X}} q_x(\mathbf{y}_{C_i}) \, d\nu(x)$$

to the density. Since $\int_{\mathcal{X}} q_x(\emptyset) \, d\nu(x) = \nu(\mathcal{X}) - \beta$, the coefficient α is

$$\sum_{n=k}^{\infty} \frac{1}{n!} (\nu(\mathcal{X}) - \beta)^{n-k} \, n(n-1) \cdots (n-k+1) = e^{\nu(\mathcal{X}) - \beta}.$$

The class of all partitions that are refinements of $D_1, \ldots, D_K$ is the Cartesian product of the sets of partitions of each D_i. Hence, for $\mathbf{y} \neq \emptyset$,

$$p(\mathbf{y}) = e^{\nu(\mathcal{X}) - \beta} \prod_{i=1}^{K} \Phi(\mathbf{y}_{D_i})$$

where

$$\Phi(\mathbf{z}) = \sum_{k \geq 1} \sum_{\mathbf{z}_{C_1}, \ldots, \mathbf{z}_{C_k}} \prod_{j=1}^{k} \int_{\mathcal{X}} q_x(\mathbf{z}_{C_j}) \, d\nu(x)$$

where $\mathbf{z}_{C_1}, \ldots, \mathbf{z}_{C_k}$ range over all unordered partitions of $\mathbf{z}$ into non-empty subconfigurations.

Since the offspring densities are hereditary excluding $\emptyset$, so is $\Phi(\cdot)$, and hence $p(\cdot)$ is hereditary. The result follows from theorem 2.3. $\square$

If the parent Poisson process is replaced by a Ripley–Kelly or connected component Markov process at range r, the cluster process described above is not connected component Markov. A heuristic explanation is that childless or 'ghost' parents may cause interaction by merging of connected sets. If we require each parent to have at least one daughter, the cluster process is connected component Markov at range $2R + r$. For a proof, see [10; 117].

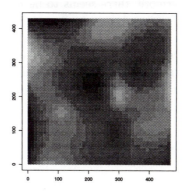

Fig. 4.7 Smoothed 2-dimensional histogram for a pattern of soil stains classified as 'vugh' in mythylene-blue coloured clay observed in a square horizontal slice of side 20 cm sampled at 4 cm depth (Hatano and Booltink, 1992). Black corresponds to high values.

Having come at the end of our review of Markov models, it should be stressed that models can be combined at libitum by simply multiplying the various interaction functions. For instance, in the next section, a hard core constraint is used in combination with an attractive area-interaction potential.

4.8 Case study

The previous sections have left us with a wide range of Markov point process models at our disposal. Here, this knowledge will be applied to analyse the vugh pattern depicted in Figure 1.2, see [87; 206; 207].

4.8.1 *Exploratory analysis*

Consider the pattern of vughs described in section 1.1. In order to gain a better understanding of these data, we will begin by computing some basic characteristics. Thus, Figure 4.7 shows the intensity of the vugh pattern, estimated by smoothing of the 2-dimensional histogram. High intensity regions are separated by empty spaces, which is typical for aggregated patterns. Furthermore, there seems to be no indication of regional inhomogeneity over the mapped region.

To make the above statements more rigorous, a simple quick test for the null hypothesis of complete spatial randomness (cf. section 1.5) divides the sampling window into m smaller cells or *quadrats*. Writing n_i for the number of observations in the ith cell, the total number of points is $n = \sum_{i=1}^{m} n_i$, and

$$\sum_{i=1}^{m} \frac{(mn_i - n)^2}{n(m-1)} \tag{4.20}$$

approximately follows a chi square distribution with $m - 1$ degrees of freedom. As a rule of thumb [46], the expected frequency n/m of points per quadrat should be at least 5. In our case, this rule suggests an arrangement of 3×3 quadrats of equal size. The counts in each cell are

5	2	4
4	14	8
7	11	5

for the vughs. The corresponding χ^2 statistic (4.20) takes the value 19.575, which is significant at the 5% level.

The quadrat count approach outlined above is particularly convenient in field work where a complete map may not be available or is too laborious to construct. However, any technique based on aggregated counts by its very nature depends on the quadrat scale. Moreover, the full map of soil stains is available to us, so that more sophisticated test statistics such as the ones introduced in section 1.8 can be used. Since plots of the (estimated) empty space and nearest-neighbour distance distribution functions are hard to interpret on their own – and indeed are usually compared to those of a binomial point process – we prefer to work with the J–function. Additionally, the latter statistic has the advantage of giving information on the range of interactions (see section 1.8). Provided they exist, plots of related statistics such as the derivative or hazard rate can also be helpful. None of the above statistics should be used for too large ranges though, as they depend on distances only, and the variance of their estimates increases with the range.

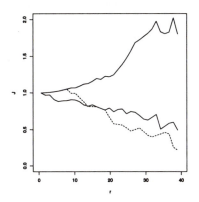

Fig. 4.8 Kaplan–Meier style estimator of the J–function for a pattern of soil stains classified as 'vugh' (dashed line) in mythylene-blue coloured clay observed in a square horizontal slice of side 20 cm sampled at 4 cm depth (Hatano and Booltink, 1992). The solid lines are upper and lower envelopes based on 19 independent simulations of a binomial point process.

Under the further assumption that the point process under consideration is isotropic, that is its distribution is invariant under rotations as well as translations, a useful statistic is the K–function defined as the expected number of points within given distance of a typical point [174]. As for the statistics based on distances, the related L–function defined by $L(r) = \sqrt{(\frac{K(r)}{\pi})}$ or the pair correlation function $g(r) = \frac{1}{2\pi r}K'(r)$ may be easier to interpret [209]. For higher order analogues, see [194; 200].

Finally, it should be emphasised that none of the statistics discussed here defines a probability distribution [12; 18].

Returning to the soil data, henceforth we assume that the process of vughs is stationary. The estimated J–function [5; 207] is given in Figure 4.8. The graph at first increases, then falls down below level 1. Thus, the pattern combines overall aggregation with inhibition at small range, probably due to the size of the stains. If we would have used the K–function instead, similar results would have been obtained [206].

To assess whether the deviations from complete spatial randomness are statistically significant, we performed a Monte Carlo test (section 1.8).

Thus, 19 independent samples from a binomial process (section 1.4, example 1.6) with the same number of points as the data were generated, and their J–functions estimated. The maximal and minimal values in the sample are recorded (Figure 4.8, solid lines) and compared to the data estimates. Since the observed values lie mostly outside of the simulation envelopes, the null hypothesis is is rejected at the 5% level.

4.8.2 *Model fitting*

Recalling that the soil stains combine overall attraction with inhibition at short range, we will try and fit an area-interaction model with a hard core

$$p(\mathbf{x}) = \alpha \, \beta^{n(\mathbf{x})} \, \gamma^{-\mu(U_{\mathbf{x}}(R_2))} \, \mathbb{I}\{||u - v|| > R_1 \text{ for all } u \neq v \in \mathbf{x}\} \qquad (4.21)$$

to the vugh pattern of Figure 1.1. Here $\beta > 0$ is the intensity parameter, R_1 is a hard core distance, $\gamma > 1$ is the attraction parameter and R_2 the radius of balls constituting $U_{\mathbf{x}}(R_2) = \bigcup_{x \in \mathbf{x}} B(x, R_2)$. For the hard core distance we took the minimum distance between data points, that is $R_1 = 3$. The interaction range is chosen by inspecting Figure 4.8; since the graph flattens down at about 40, $R_2 = 20$. Alternatively, a profile likelihood approach could have been taken [13].

To fit (4.21), the method described in section 3.7 was used. Note that the sufficient statistics are the cardinality $N(\mathcal{X})$ and the covered area $\mu(U_X(R_2))$. Thus, the Monte Carlo log likelihood ratio with respect to a reference parameter vector $\theta_0 = (\beta_0, \log \gamma_0)$ is given by

$$n(\mathbf{x}) \log \left(\frac{\beta}{\beta_0} \right) - \mu(U_{\mathbf{x}}(R_2)) \{ \log \gamma - \log \gamma_0 \}$$

$$-\log \left\{ \frac{1}{n} \sum_{i=1}^{n} \left(\frac{\beta}{\beta_0} \right)^{N(X_i)} \exp \left[-\mu(U_{X_i}(R_2)) \{ \log \gamma - \log \gamma_0 \} \right] \right\} \qquad (4.22)$$

where $\mathbf{x}$ is the data pattern and X_i, $i = 1, \ldots, n$, are successive states in a Metropolis–Hastings algorithm (cf. section 3.2). In order to alleviate edge effects, we sampled point patterns in the observation window dilated by a ball of radius $2R_2$, before clipping to the original window. Independent realisations over a variety of parameter values were generated and the mean values of the sufficient statistics computed. After a few such trial runs,

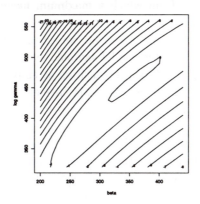

Fig. 4.9 Contour plot of the log likelihood with reference parameter $\beta_0 = 320$, $\log \gamma_0 = 440$.

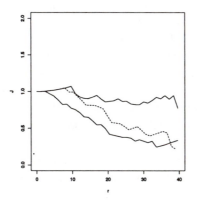

Fig. 4.10 Kaplan–Meier style estimator of the J–function for a pattern of soil stains classified as 'vugh' (dashed line) in mythylene-blue coloured clay observed in a square horizontal slice of side 20 cm sampled at 4 cm depth (Hatano and Booltink, 1992). The solid lines are upper and lower envelopes based on 19 simulations of the fitted area-interaction with hard core model.

we zoomed in on $\beta_0 = 320$ and $\log \gamma_0 = 440$ and ran 20000 steps of the Metropolis–Hastings algorithm. The contour plot of the log likelihood is depicted in Figure 4.9, from which a maximum likelihood estimate $\hat{\theta} = (354, 461)$ is obtained.

Finally, to assess the goodness of fit, we estimated the J–function for 19 simulations from the fitted model sampled every 5000 steps in a long run of 115000 Metropolis–Hastings transition attempts, allowing a burn-in period of 20000 steps. The upper and lower envelopes are given by the solid lines in Figure 4.10, and should be compared to the estimated J–function of the data pattern (the dashed line). All graphs coincide up to the hard core distance; at somewhat larger ranges, the data is at the top end of the graph, but soon lies in between the upper and lower envelopes. Hence, the fitted model cannot be rejected. It is interesting to compare Figure 4.10 to the corresponding one for a binomial null hypothesis (cf. Figure 4.8).

4.9   Interpolation and extrapolation

The aim of this section is to illustrate how marked point processes may be used in the interpretation of spatial data. For instance in telecommunication, marked point processes are a useful tool in computing network costs [57]. Also in object recognition problems [2; 4; 7; 72; 73; 74; 127; 146; 177; 182; 183; 189; 164], a scene is conveniently represented by a point process of locations marked by size and shape parameters. Inference may be based on the probability distribution $p(\mathbf{y} \mid \mathbf{x})$ of the observed image $\mathbf{y}$ parametrised by the 'true' pattern $\mathbf{x}$ but, since maximum likelihood solutions tend to suffer from multiple response, a Bayesian approach maximising a posterior optimality criterion [58; 188] seems preferable. Indeed, a repulsive Markov marked point process as prior distribution has proved effective in combatting overfitting [7; 189; 127]. A natural neighbourhood relation in the present context is the overlapping object relation. From a computational point of view, an important advantage of using Markov priors is that the posterior likelihood ratios for addition, deletion of modifying an object (the building blocks of the samplers discussed in Chapter 3) are easy to compute and moreover depend on a neighbourhood of the modified object only.

Yet another class of applications concern situations were one is interested in predicting features of a point process beyond the window of observation, or in interpolation of unobservable aspects of the pattern. In both cases, a stochastic model must be fitted to the data, and relevant properties of the conditional distribution given the observations are computed in order to perform the prediction.

Here, we will illustrate some typical issues by means of the problem of identifying clusters in a spatial point pattern [7; 9; 114; 117; 121]. The issue arises in several contexts, from detecting clusters in the epidemiology of rare diseases to grouping features when searching large image data bases on the world wide web.

Most traditional clustering approaches build a tree based on some similarity measure [30]. From this tree, the number of clusters and the corresponding partition is decided in an ad hoc (and mostly subjective) manner. More recently, model based clustering techniques [42; 44] consider finite mixture models. The number of groups is determined by a Bayes factors or AIC criterion, and given the number of mixture components, model parameters are estimated by maximum likelihood, often using a variant of the EM-algorithm [43]. Most applications also allow a 'don't know' class for outliers or noise. The cluster centers only play an implicit role – approximated by the center of gravity, principal axis or other 'mean' of the detected clusters – if they appear at all.

The advantage of taking a point process approach is that the number of clusters, their centres and the offspring labelling are treated simultaneously. Moreover, in contrast to tree-based techniques, a full posterior distribution is sampled, thus facilitating the estimation of conditional expectations of quantities of interest. Finally note the problem involves both interpolation of the cluster centers and sibling partitioning, as well as extrapolation beyond the observation window.

4.9.1 *Model*

Figure 1.10 depicts a realisation of a Matérn cluster process [129]. In the remainder of this section, we will analyse this pattern by detecting the clusters, locating their centres, and assigning each data point to a cluster.

From section 1.6, recall that in a Matérn cluster process a parent at $x \in \mathcal{X}$ is replaced by a Poisson number of offspring positioned independently and uniformly in a ball centred at x. In other words, the daughters Z_x of

x form an inhomogeneous Poisson process with intensity function

$$h(y \mid x) = \begin{cases} \mu & \text{if } d(x,y) \leq R \\ 0 & \text{otherwise} \end{cases} \qquad (4.23)$$

where $\mu > 0$ is the intensity parameter and $R > 0$ the scatter radius. If the point process Z_0 of outliers is independent of all other offspring, and modelled by a Poisson process of intensity $\epsilon > 0$ [42; 117; 121] then, conditionally on a parent configuration $X = \mathbf{x}$, by theorem 1.4 the total pattern observed in a window A is a Poisson process with intensity function

$$\lambda(y \mid \mathbf{x}) = \epsilon + \sum_{x \in \mathbf{x}} h(y \mid x), \qquad (4.24)$$

$y \in A$. In Figure 1.10, the daughters were observed within the window $A = [0,1]^2$, and we may try and reconstruct the parent process within some compact set $\mathcal{X} \supseteq A \oplus B(0,R)$. Although in the sequel we shall focus on (4.23) for concreteness' sake, it should be clear that the approach outlined below remains valid for other intensity functions [9; 117; 121].

To estimate the clusters, consider the conditional distribution given the data of the marked point process W of parents labelled by their offspring in A. An arbitrary dummy point x_0 will serve as parent to the configuration Z_0 of outliers. Now, if $\mathbf{u}$ is the union of all daughter points and outliers observed in A, for any partition $\mathbf{z}_i$ of $\mathbf{u}$, W has conditional density [117; 121]

$$p_{W \mid U}((x_i, \mathbf{z}_i)_{i \leq n} \mid \mathbf{u}) =$$
$$\mathbf{P}(\mathbf{z}_0, \ldots, \mathbf{z}_n \mid x_0, \ldots, x_n, \mathbf{u})\, p_{X \mid U}(\mathbf{x} \mid \mathbf{u}) =$$
$$c(\mathbf{u})\, \mathbf{P}(\mathbf{z}_0, \ldots, \mathbf{z}_n \mid x_0, \ldots, x_n, \mathbf{u})\, p_{U \mid X}(\mathbf{u} \mid \mathbf{x})\, p_X(\mathbf{x}),$$

with respect to the distribution of a unit rate Poisson process on $\mathcal{X}$ marked by a partition of $\mathbf{u}$. Here $\mathbf{P}(\mathbf{z}_0, \ldots, \mathbf{z}_n \mid \mathbf{x}, \mathbf{u})$ denotes the conditional probability that those points of $\mathbf{u}$ in the subset $\mathbf{z}_i$ are ascribed to x_i, $p_{X \mid U}(\cdot \mid \mathbf{u})$ is the conditional density with respect to a unit rate Poisson process on $\mathcal{X}$ of the union of daughter and noise points, and $p_X(\cdot)$ is the (prior) density for the parent process X with respect to a unit rate Poisson process on $\mathcal{X}$. Finally, $c(\mathbf{u})$ is a normalising constant depending on $\mathbf{u}$.

Since conditionally on $\mathbf{x}$ the total offspring including that of the dummy

x_0 form a Poisson process with intensity function given by (4.24),

$$p_{U|X}(\mathbf{u} \mid \mathbf{x}) = \exp\left[\int_A (1 - \lambda(a \mid \mathbf{x})) \, da\right] \prod_{u \in \mathbf{u}} \lambda(u \mid \mathbf{x}). \qquad (4.25)$$

Provided $\mathbf{x}$ and $\mathbf{u}$ are compatible in that there exists at least one partition $(\mathbf{z}_0', \dots, \mathbf{z}_n')$ of $\mathbf{u}$ for which $\prod_{i=0}^n \prod_{a \in \mathbf{z}_i'} h(a \mid x_i) > 0$, the partition probabilities are

$$\mathbf{P}(\mathbf{z}_0, \dots, \mathbf{z}_n \mid \mathbf{x}, \mathbf{u}) = \frac{\prod_{i=0}^n \prod_{a \in \mathbf{z}_i} h(a \mid x_i)}{\prod_{j=1}^m \lambda(u_j \mid \mathbf{x})},$$

coding $h(\cdot \mid x_0) \equiv \epsilon$ for the noise component [117; 121]. In terms of the function $\varphi : \mathbf{u} \mapsto \{0, 1, \dots, n\}$ allocating each data point to its parent, equation (4.26) implies that the points are ascribed to a cluster center *independently* with probabilities [117; 121]

$$\mathbf{P}(\varphi(u) = i \mid \mathbf{x}, \mathbf{u}) = \frac{h(u \mid x_i)}{\lambda(u \mid \mathbf{x})}, \qquad i = 0, \dots, n. \qquad (4.26)$$

Finally, the choice of the prior term $p_X(\cdot)$ is up to the discretion of the analyst. Of course, a Poisson process may be taken, but it seems more natural to incorporate some repulsion to avoid 'overfitting' [117; 121]. Thus, we choose a hard core process with density given by (2.1).

4.9.2 *Posterior sampling*

Because of the very simple form of the cluster allocation labelling (4.26), sampling from $P_{W|U}(\cdot \mid \mathbf{u})$ can be performed sequentially: first generate the parent configuration, then ascribe the daughters to a cluster independently according to (4.26). Here we take the spatial birth-and-death approach discussed in section 3.4. Following example 3.7, one may take a constant death rate and birth rate proportional to the posterior conditional intensity

$$\lambda_{X|U}(\xi; \mathbf{x}) = \lambda_X(\xi; \mathbf{x}) \exp\left[-\int_A h(a \mid \xi) \, da\right] \prod_{u \in \mathbf{u}} \left[1 + \frac{h(u \mid \xi)}{\lambda(u \mid \mathbf{x})}\right].$$

However, the total birth rate would be difficult to compute, and the product over data points in the expression above may be very large. For these reasons, we prefer to work with

$$b(\mathbf{x}, \xi) = \lambda_X(\xi; \mathbf{x}) \left[1 + \sum_{u \in \mathbf{u}} \frac{h(u \mid \xi)}{\epsilon} \right] \tag{4.27}$$

which is less peaked than the posterior conditional intensity, while retaining the desirable property of placing most new-born points in the vicinity of points of $\mathbf{u}$. Since the hard core prior distribution is locally stable with the intensity parameter λ as upper bound on the Papangelou conditional intensity, the total birth rate

$$B(\mathbf{x}) = \int_X b(\mathbf{x}, \xi) \, d\xi \leq \lambda \left[\mu(\mathcal{X}) + \frac{1}{\epsilon} \sum_{u \in \mathbf{u}} \int_X h(u \mid \xi) \, d\xi \right] := G$$

is bounded above by a constant G that is easy to evaluate. By the detailed balance equations (3.5), the death rate for deleting ξ from configuration $\mathbf{x} \cup \{\xi\}$ is

$$d(\mathbf{x} \cup \{\xi\}, \xi) = \frac{\exp \left[\int_A h(a \mid \xi) \, d\mu(a) \right]}{\prod_{u \in \mathbf{u}} \left[1 + \frac{h(u \mid \xi)}{\lambda(u \mid \mathbf{x})} \right]} \left[1 + \sum_{u \in \mathbf{u}} \frac{h(u \mid \xi)}{\epsilon} \right]. \tag{4.28}$$

The total death rate from $\mathbf{x}$ is denoted by $D(\mathbf{x})$.

Note that the birth and death rates (4.27)–(4.28) only depend locally on $\mathbf{x}$, a property that is very important from a computational point of view.

In summary, if the current state is $\mathbf{x}$, after an exponentially distributed sojourn time of rate $G + D(\mathbf{x})$, with probability $D(\mathbf{x})/(G + D(\mathbf{x}))$ a point of $\mathbf{x}$ is deleted according to the distribution $d(\mathbf{x}, x_i)/D(\mathbf{x})$; a birth is proposed with the complementary probability $G/(G + D(\mathbf{x}))$ by sampling a candidate ξ from the mixture density $\frac{\lambda}{G} \left[1 + \sum_{u \in \mathbf{u}} \frac{h(u \mid \xi)}{\epsilon} \right]$, which is then accepted with probability $\lambda_X(\xi; \mathbf{x})/\lambda$.

Lemma 4.2 *There exists a unique spatial birth-and-death process with transition rates given by (4.27) and (4.28). It has unique equilibrium distribution $p_{X \mid U}(\cdot \mid \mathbf{u})$, to which it converges in distribution from any initial state.*

The above lemma holds for any locally stable Markov prior distribution, and any uniformly bounded intensity function [9].

Proof.   Note that $h(\cdot \mid \cdot) \leq \mu$ uniformly in both its arguments. Hence the total birth rate is uniformly bounded from above by $\lambda \mu(\mathcal{X}) \left(1 + m\frac{\mu}{\epsilon}\right)$ if $\mathbf{u}$ contains m points. Furthermore, $d(\mathbf{x}, x_i) \geq \left(1 + \frac{\mu}{\epsilon}\right)^{-m} = C > 0$. Hence the total death rate $D(\mathbf{x})$ is bounded from below by $n(\mathbf{x})\, C$. By theorem 3.3, the result follows. $\square$

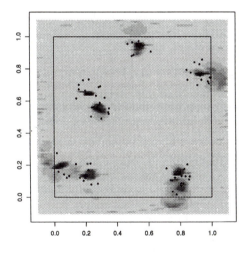

Fig. 4.11   The points are those of a realisation of a Matérn process on $[0,1]^2$ with parent intensity $\lambda = 10$, mean number of daughters $\nu = 10$ per cluster and radius $r = 0.1$. The image depicts the smoothed posterior parent frequencies on logarithmic scale estimated over 250 time units of the spatial birth and death process described in the text, initialised by an exact sample from the posterior distribution (black corresponding to high probability). The prior distribution for parents was a hard core model with distance 0.05, the outlier intensity $\epsilon = 1.0$.

4.9.3   *Monotonicity properties and coupling from the past*

The spatial birth-and-death process considered in the previous subsection can be made exact using coupling from the past.

Lemma 4.3   *The posterior distribution specified by $p_{X|U}(\cdot \mid \mathbf{u})$ is hereditary and repulsive. Furthermore, the birth and death rates defined in (4.27)–(4.28) respect the inclusion ordering in the sense that if $\mathbf{x} \subseteq \mathbf{x}'$ then $b(\mathbf{x}, \xi) \geq b(\mathbf{x}', \xi)$ for all $\xi \in \mathcal{X}$, while $d(\mathbf{x}, x_i) \leq d(\mathbf{x}', x_i)$ for $x_i \in \mathbf{x}$.*

Note that the above lemma remains valid if we replace our hard core prior by any repulsive Markov point process [9].

Proof.   Since $p_{U|X}(\cdot) > 0$, the posterior density is hereditary. Let $\mathbf{x} \subseteq \mathbf{x}'$. Then

$$\prod_{u \in \mathbf{u}} \left[1 + \frac{h(u \mid \xi)}{\epsilon + \sum_{i=1}^{n(\mathbf{x})} h(u \mid x_i)} \right] \geq \prod_{u \in \mathbf{u}} \left[1 + \frac{h(u \mid \xi)}{\epsilon + \sum_{i=1}^{n(\mathbf{x}')} h(u \mid x_i)} \right]. \quad (4.29)$$

Since the hard core process is repulsive, $\lambda_X(\xi; \mathbf{x}) \geq \lambda_X(\xi; \mathbf{x}')$, and consequently

$$\begin{aligned}
\lambda_{X|U}(\xi; \mathbf{x}) &= \lambda_X(\xi; \mathbf{x}) \exp\left[-\int_A h(a \mid \xi)\, da\right] \prod_{u \in \mathbf{u}} \left[1 + \frac{h(u \mid \xi)}{\epsilon + \sum_{i=1}^{n(\mathbf{x})} h(u \mid x_i)} \right] \\
&\geq \lambda_{X|U}(\xi; \mathbf{x}').
\end{aligned}$$

The ordering between the birth rates follows directly from the fact that the hard core process is repulsive. The statement about the death rates follows from (4.29).   □

In order to define a coupling from the past algorithm, note that by lemma 4.3 and the fact that the posterior conditional intensity at ξ is bounded by $\lambda \prod_{u \in \mathbf{u}} \left(1 + \frac{h(u|\xi)}{\epsilon}\right)$, algorithm 3.4 is applicable. However, the upper bound is far from tight, resulting in a slow convergence.

To describe a taylor-made coupling from the past algorithm, return to the transition rates (4.27)–(4.28). Now

$$b(\mathbf{x}, \xi) \leq \lambda \left[1 + \sum_{u \in \mathbf{u}} \frac{h(u \mid \xi)}{\epsilon} \right] = \bar{b}(\xi). \quad (4.30)$$

Regarding the death rate, note that

$$1 + \frac{h(u \mid \xi)}{\lambda(u \mid \mathbf{x})} \leq 1 + \frac{\mu}{\epsilon + \mu} \leq 2$$

if $u_j \in B(\xi, R) \cap U_\mathbf{x}$, where as usual $U_\mathbf{x} = \cup_{x_i \in \mathbf{x}} B(x_i, R)$ denotes the union of balls centred at the points of $\mathbf{x}$. It follows that the death rate $d(\mathbf{x} \cup \{\xi\}, \xi)$ is minorised by

$$\underline{d}(\mathbf{x} \cup \{\xi\}, \xi) = d(\mathbf{x} \cup \{\xi\}, \xi) \prod_{j:u_j \in B(\xi, R) \cap U_\mathbf{x}} \left(\frac{1}{2} + \frac{\mu}{2\,\lambda(u_j \mid \mathbf{x})} \right). \quad (4.31)$$

The equilibrium distribution $\pi(\cdot)$ of the spatial birth and death process given by (4.30)–(4.31) can be found by solving the detailed balance equations

$$\pi(\mathbf{x})\,\bar{b}(\xi) = \pi(\mathbf{x} \cup \{\xi\})\,\underline{d}(\mathbf{x} \cup \{\xi\}, \xi)$$

yielding

$$\pi(\mathbf{x}) \propto \gamma^{-n(\mathbf{u} \cap U_\mathbf{x})} \prod_{i=1}^{n(\mathbf{x})} \beta(x_i), \quad (4.32)$$

a generalised area-interaction process with intensity function

$$\beta(\xi) = \lambda \exp\left[-\int_A h(a \mid \xi)\, da \right] 2^{n(\mathbf{u} \cap B(\xi, R))}$$

and interaction parameter $\gamma = (\frac{1}{2} + \frac{\mu}{2\epsilon})^{-1}$. A sample may be obtained using the representation in terms of two Poisson processes discussed in example 4.7 of section 4.3.

We are now able to formulate a coupling from the past algorithm [9].

Algorithm 4.2   Set $T = 1$ and let $D(0)$ be a sample from the generalised area-interaction process (4.32). Repeat

- extend $D(\cdot)$ backwards until time $-T$ by means of a spatial birth-and-death process with birth rate $\bar{b}(\cdot)$ and death rate $\underline{d}(\cdot, \cdot)$;
- generate a lower bound process $L_{-T}(\cdot)$ and an upper bound process $U_{-T}(\cdot)$ on $[-T, 0]$ as described below;
- if $U_{-T}(0) = L_{-T}(0)$, return the common state $U_{-T}(0)$; otherwise set $T := 2 * T$;

until the upper and lower processes have coalesced.

The minimal and maximal processes $L_{-T}(\cdot)$ and $U_{-T}(\cdot)$ are coupled to the dominating process $D(\cdot)$ as follows. The lower process is initialised with the empty set, $L_{-T}(-T) = \emptyset$; the maximal process is initialised with the value of the dominating process at time $-T$: $U_{-T}(-T) = D(-T)$. To each forward transition time $t \in (-T, 0]$ of $D(\cdot)$ corresponds an update of $L_{-T}(\cdot)$ and $U_{-T}(\cdot)$. In case of a death (i.e. a backwards birth), the transition is mimicked in both $U_{-T}(\cdot)$ and $L_{-T}(\cdot)$. In case of a birth $D(t) = D(t-) \cup \xi$, the point ξ is added to $U_{-T}(t-)$ with probability

$$\frac{b(L_{-T}(t-), \xi) \, \underline{d}(L_{-T}(t-) \cup \{\xi\})}{\bar{b}(\xi) \, d(L_{-T}(t-) \cup \{\xi\})}.$$

The above expression with $U_{-T}(t-)$ replacing $L_{-T}(t-)$ holds for the acceptance probability of the birth transition in the lower process. All random numbers generated previously are reused.

To see the validity of algorithm 4.2, write r_{hc} for the hard core radius. Then

$$\frac{b(\mathbf{x}, \xi) \, \underline{d}(\mathbf{x} \cup \{\xi\})}{\bar{b}(\xi) \, d(\mathbf{x} \cup \{\xi\})} = \mathbb{I}\{d(\xi, \mathbf{x}) > r_{hc}\} \prod_{j : u_j \in B(\xi, R) \cap U_\mathbf{x}} \left(\frac{1}{2} + \frac{\mu}{2\lambda(u_j | \mathbf{x})}\right)$$

is decreasing in $\mathbf{x}$, so the inclusion order is preserved at each transition. By arguments analogous to those for algortihm 3.4, it now follows that Algorithm 4.2 almost surely terminates and outputs an unbiased sample from $p_{X|U}(\cdot \mid \cdot)$.

As an illustration, we ran algorithm 4.2 for the data of Figure 1.10 with a hard core prior distribution to ensure that two clusters are separated by at least a distance 0.10 and an outlier intensity $\epsilon = 1.0$. For these values, coupling occurred already after 2 time units. Rather than give a single realisation of the posterior parent distribution, we computed the frequency with which a parent was obtained at each location in a long run of the spatial birth-and-death process specified by (4.27)–(4.28) over 250 time units. The sampler was initialised by the pattern outputted by algorithm 4.2, so that at each time, the state of the sampler has the required distribution. Finally, note that if the parameters had been unknown, they could have been estimated easily by means of the EM-algorithm [43].

Bibliography

[1] M.P. Allen and D.J. Tildesley. *Computer simulation of liquids.* Oxford University Press, Oxford, 1987.

[2] Y. Amit, U. Grenander, and M. Piccioni. Structural image restoration through deformable templates. *Journal of the American Statistical Association*, 86:376–387, 1991.

[3] G. Ayala and A. Simó. Bivariate random closed sets and nerve fibre degeneration. *Advances in Applied Probability (SGSA)*, 27:293–305, 1995.

[4] R.G. Aykroyd and P.J. Green. Global and local priors, and the location of lesions using gamma camera imagery. *Philosophical Transactions of the Royal Society of London, Series A*, 337:323–342, 1991.

[5] A.J. Baddeley and R.D. Gill. Kaplan-Meier estimators for interpoint distance distributions of spatial point processes. *Annals of Statistics*, 25:263–292, 1997.

[6] A.J. Baddeley and M.N.M. van Lieshout. ICM for object recognition. In Y. Dodge and J. Whittaker, editors, *Computational statistics*, volume 2, pages 271–286, Heidelberg-New York, 1992. Physica/Springer.

[7] A.J. Baddeley and M.N.M. van Lieshout. Stochastic geometry models in high-level vision. In K.V. Mardia and G.K. Kanji, editors, *Statistics and Images Volume 1, Advances in Applied Statistics, a supplement to Journal of Applied Statistics Volume 20*, pages 231–256, Abingdon, 1993. Carfax.

[8] A.J. Baddeley and M.N.M. van Lieshout. Area-interaction point processes. *Annals of the Institute of Statistical Mathematics*, 47:601–619, 1995.

[9] A.J. Baddeley, M.N.M. van Lieshout, and et al. Extrapolating and interpolating spatial patterns. In preparation.

[10] A.J. Baddeley, M.N.M. van Lieshout, and J. Møller. Markov properties of cluster processes. *Advances in Applied Probability*, 28:346–355, 1996.

[11] A.J. Baddeley and J. Møller. Nearest-neighbour Markov point processes and random ooto. *International Statistical Review*, 57:89–121, 1989.

[12] A.J. Baddeley and B.W. Silverman. A cautionary example for the use of second-order methods for analysing point patterns. *Biometrics*, 40:1089–1094, 1984.

[13] A.J. Baddeley and T.R. Turner. Practical maximum pseudolikelihood for spatial point patterns. Technical Report 98-16, University of Western Australia, 1998.

[14] A.A. Barker. Monte Carlo calculation of the radial distribution functions for a proton-electron plasma. *Australian Journal of Physics*, 18:119–133, 1965.

[15] V. Barnett, R. Payne, and V. Steiner. *Agricultural sustainability: economic, environmental and statistical considerations*. Wiley, Chichester, 1995.

[16] M.S. Bartlett. *The statistical analysis of spatial pattern*. Chapman and Hall, London, 1975.

[17] H. Bauer. *Probability theory and elements of measure theory*. Holt, Rinehart and Winston, New York, 1972.

[18] T. Bedford and J. van den Berg. A remark on Van Lieshout and Baddeley's J-function for point processes. *Advances in Applied Probability (SGSA)*, 29:19–25, 1997.

[19] M. Berman and T.R. Turner. Approximating point process likelihoods with GLIM. *Applied Statistics*, 41:31–38, 1992.

[20] J. Besag. Some methods of statistical analysis for spatial data. *Bulletin of the International Statistical Institute*, 47:77–92, 1977.

[21] J. Besag and P.J. Diggle. Simple Monte Carlo tests for spatial pattern. *Applied Statistics*, 26:327–333, 1977.

[22] J. Besag and P.J. Green. Spatial statistics and Bayesian computa-

tion. *Journal of the Royal Statistical Society, Series B*, 36:25–38, 1993.

[23] J. Besag, P.J. Green, D. Higdon, and K. Mengersen. Bayesian computation and stochastic systems. *Statistical Science*, 10:3–41, 1995.

[24] J. Besag, R. Milne, and S. Zachary. Point process limits of lattice processes. *Journal of Applied Probability*, 19:210–216, 1982.

[25] K. Binder and D.W. Heermann. *Monte Carlo methods in statistical physics*. Springer, Berlin, 1986.

[26] J. Bouma, A. Jongerius, O. Boersma, A. de Jager, and D. Schoonderbeek. The function of different types of macropores during saturated flow through swelling soil horizons. *Journal of the American Soil Science Society*, 41:945–950, 1977.

[27] J. Bronowski and J. Neyman. The variance of the measure of a two-dimensional random set. *Annals of Mathematical Statistics*, 16:330–341, 1945.

[28] H. Cai. Cluster algorithms for spatial point processes with symmetric and shift invariant interactions. *Journal of Computational and Graphical Statistics*, 8:353–372, 1999.

[29] N.R. Campbell. The study of discontinuous phenomena. *Proceedings of the Cambridge Philosophical Society*, 15:117–136, 1909.

[30] C. Chatfield and A.J. Collins. *Introduction to multivariate analysis*. Chapman and Hall, London, 1980.

[31] J.T. Chayes, L. Chayes, and R. Kotecky. The analysis of the Widom–Rowlinson model by stochastic geometric methods. *Communications in Mathematical Physics*, 172:551–569, 1995.

[32] L. Chayes and J. Machta. Graphical representations and cluster algorithms part II. *Physica A*, 239:542, 1997.

[33] F. Chen and P.A. P.A. Kelly. Algorithms for generating and segmenting morphologically smooth binary images. In *Proceedings of the 26th Conference on Information Sciences*, page 902, 1992.

[34] Y.C. Chin and A.J. Baddeley. On connected component Markov point processes. *Advances in Applied Probability (SGSA)*, 31:279–282, 1999.

[35] S.N. Chiu and D. Stoyan. Estimators of distance distributions for spatial patterns. *Statistica Neerlandica*, 52:239–246, 1998.

[36] G. Choquet. Theory of capacities. *Annals of the Fourier Institute*, V:131–295, 1953/1954.

[37] K.L. Chung. *Markov chains with stationary transition probabilities.* Springer, Berlin, 2nd edition, 1967.

[38] P. Clifford and G. Nicholls. Comparison of birth-and-death and Metropolis–Hastings Markov chain Monte Carlo for the Strauss process, 1994. Manuscript, Oxford University.

[39] D.R. Cox and V. Isham. *Point processes.* Chapman and Hall, London, 1980.

[40] N.A.C. Cressie. *Statistics for spatial data.* Wiley, New York, 1991.

[41] D. Daley, D.J.and Vere-Jones. *An introduction to the theory of point processes.* Springer, New York, 1988.

[42] A. Dasgupta and A.E. Raftery. Detecting features in spatial point processes with clutter via model-based clustering. *Journal of the Americal Statistical Association,* 93:294–302, 1998.

[43] A.P. Dempster, N.M. Laird, and D.B. Rubin. Maximum likelihood from incomplete data via the EM algorithm. *Journal of the Royal Statistical Society, Series B,* 39:1–37, 1977.

[44] J. Diebolt and C.P. Robert. Estimation of finite mixture distributions through Bayesian sampling. *Journal of the Royal Statistical Society, Series B,* 56:363–375, 1994.

[45] P.J. Diggle. On parameter estimation and goodness-of-fit testing for spatial point processes. *Biometrics,* 35:87–101, 1979.

[46] P.J. Diggle. *Statistical analysis of spatial point patterns.* Academic Press, London, 1983.

[47] P.J. Diggle. A kernel method for smoothing point process data. *Applied Statistics,* 34:138–147, 1985.

[48] P.J. Diggle, T. Fiksel, P. Grabarnik, Y. Ogata, D. Stoyan, and M. Tanemura. On parameter estimation for pairwise interaction processes. *International Statistical Review,* 62:99–117, 1994.

[49] P.J Diggle, D.J. Gates, and A. Stibbard. A nonparametric estimator for pairwise interaction point processes. *Biometrika,* 74:763–770, 1987.

[50] P.J. Diggle and J.S. Marron. Equivalence of smoothing parameter selection in density and intensity estimation. *Journal of the American Statistical Association,* 83:793–800, 1988.

[51] P. Droogers. *Quantifying differences in soil structure induced by farm management.* PhD thesis, Wageningen Agricultural University, 1997.

[52] R.G. Edwards and A.D. Sokal. Generalization of the Fortuin–Kasteleyn–Swndsen–Wang representation and Monte Carlo algorithm. *Physical Reviews, Series D*, 38:2009–2012, 1988.

[53] T. Fiksel. Estimation of parametrized pair potentials of marked and non-marked Gibbsian point processes. *Elektronische Informationsverarbeitung und Kybernetik*, 20:270–278, 1984.

[54] T. Fiksel. Estimation of interaction potentials of Gibbsian point processes. *Statistics*, 19:77–86, 1988.

[55] J.A. Fill. An interruptible algorithm for pefect sampling via Markov chains. *Annals of Applied Probability*, 8:131–162, 1998.

[56] G.S. Fishman. *Monte Carlo, concepts, algorithms and applications*. Springer, New York, 1996.

[57] A. Frey and V. Schmidt. Marked point processes in the plane - a survey with applications to spatial modelling of telecommunications networks. *Advances in Performance Analysis*, 2, 1998.

[58] A. Frigessi and H. Rue. Bayesian image classification with Baddeley's delta loss. Technical Report 5/1995, University of Trondheim, 1995.

[59] F. Garwood. The variance of the overlap of geometrical figures with reference to a bombing problem. *Biometrika*, 34:1–17, 1947.

[60] V. Gavrikov and D. Stoyan. The use of marked point processes in ecological and environmental forest studies. *Environmental and ecological statistics*, 2:331–344, 1995.

[61] C.J. Geyer. Markov chain Monte Carlo maximum likelihood. In E.M. Keramidas, editor, *Computer science and statistics: Proceedings of the 23rd symposium on the Interface*, pages 156–163, Fairfax Station, 1991. Interface Foundation.

[62] C.J. Geyer. On the convergence of Monte Carlo maximum likelihood calculations. *Journal of the Royal Statistical Society, Series B*, 56:261–274, 1994.

[63] C.J. Geyer. Likelihood inference for spatial point processes. In O. Barndorff-Nielsen, W.S. Kendall, and M.N.M. van Lieshout, editors, *Stochastic geometry, likelihood, and computation*, London, 1999. Chapman and Hall.

[64] C.J. Geyer and J. Møller. Simulation procedures and likelihood inference for spatial point processes. *Scandinavian Journal of Statistics*, 21:359–373, 1994.

[65] C.J. Geyer and E.A. Thompson. Constrained Monte Carlo maximum likelihood for dependent data. *Journal of the Royal Statistical Society, Series B*, 54:657–699, 1992.

[66] W.R. Gilks, S. Richardson, and D.J. Spiegelhalter. *Markov chain Monte Carlo in practice*. Chapman and Hall, London, 1996.

[67] E. Glötzl. Bemerkungen zu einer Arbeit von O.K. Kozlov. *Mathematische Nachrichten*, 94:277–289, 1980.

[68] E. Glötzl. Lokale Energien und Potentiale für Punktprozesse. *Mathematische Nachrichten*, 96:195–206, 1980.

[69] M. Goulard, P. Grabarnik, and A. Särkkä. Parameter estimation for marked Gibbs point processes through the maximum pseudo-likelihood method. *Scandinavian Jouranl of Statistics*, 23:365–379, 1996.

[70] J. Goutsias and K. Sivakumar. Multiresolution morphological stochastic image modeling. *CWI Quarterly*, 11:347–369, 1998.

[71] P.J. Green. Reversible jump MCMC computation and Bayesian model determination. *Biometrika*, 82:711–732, 1995.

[72] U. Grenander. Tutorial in pattern theory. Technical report, Brown University, Rhode Island, 1983.

[73] U. Grenander and D.M. Keenan. A computer experiment in pattern theory. *Communications in Statistics - Stochastic Models*, 5:531–553, 1989.

[74] U. Grenander and M.I. Miller. Representations of knowledge in complex systems. *Journal of the Royal Statistical Society, Series B*, 56:549–603, 1994.

[75] G.R. Grimmett and D.R. Stirzaker. *Probability and random processes*. Clarendon Press, Oxford, 1992.

[76] H. Haario and E. Saksman. Simulated annealing process in general state space. *Advances in Applied Probability*, 23:886–893, 1991.

[77] O. Häggström, M.N.M. van Lieshout, and J. Møller. Characterisation and simulation results including exact simulation for some spatial point processes. *Bernoulli*, 5:641–659, 1999.

[78] O. Häggström and K. Nelander. Exact sampling for anti-monotone systems. *Statistica Neerlandica*, 52:360–380, 1998.

[79] P.R. Halmos. *Measure theory*. Springer, New York, 1974.

[80] J.M. Hammersley and D.C. Handscomb. *Monte Carlo methods*. Methuen, London, 1964.

[81] J.M. Hammersley, J.W.E. Lewis, and J.S. Rowlinson. Relationships between the multinomial and Poisson models of stochastic processes, and between the canonical and grand canonical ensembles in statistical mechanics, with illustrations and Monte Carlo methods for the penetrable sphere model of liquid-vapour equilibrium. *Sankhya, Series A*, 37:457–491, 1975.

[82] K.-H. Hanisch. *Beiträge zur Stereologie und Geometriestatistik unter besonderer Berücksichtigung von Momentmasen.* PhD thesis, Bergakademie Freiberg, 1983.

[83] K.-H. Hanisch. Some remarks on estimators of the distribution function of nearest neighbour distance in stationary spatial point processes. *Mathematische Operationsforschung und Statistik, Series Statistik*, 15:409–412, 1984.

[84] M.B. Hansen, R.D. Gill, and A.J. Baddeley. Kaplan-Meier type estimators for linear contact distributions. *Scandinavian Journal of Statistics*, 23:129–155, 1996.

[85] R.D. Harkness and V.S. Isham. A bivariate spatial point pattern of ants' nests. *Applied Statistics*, 32:293–303, 1983.

[86] W. Hastings. Monte Carlo sampling methods using Markov chains and their application. *Biometrika*, 57:97–109, 1970.

[87] R. Hatano and H.W.G. Booltink. Using fractal dimensions of stained flow patterns in a clay soil to predict bypass flow. *Journal of Hydrology*, 135:121–131, 1992.

[88] H.J.A.M. Heijmans. *Morphological image operators.* Academic Press, Boston, 1994.

[89] J. Heikkinen and E. Arjas. Non-parametric Bayesian estimation of a spatial Poisson intensity. *Scandinavian Journal of Statistics*, 25:435–450, 1998.

[90] J. Heikkinen and A. Penttinen. Bayesian smoothing in estimation of the pair potential function of Gibbs point processes. Technical Report 17, University of Jyväskylä, 1995.

[91] P.C.T. van der Hoeven. *On point processes.* Mathematisch Centrum Tract 165, Amsterdam, 1983.

[92] F. Huang and Y. Ogata. Improvements of the maximum pseudo-likelihood estimators in various spatial statistical models. *Journal of Computational and Graphical Statistics*, 8:510–530, 1999.

[93] V.I. Isham. Multitype Markov point processes: some approxi-

mations. *Proceedings of the Royal Society of London, Series A*, 391:39–53, 1984.

[94] E.B. Vedel Jensen and L. Stougaard Nielsen. Inhomogeneous Markov point processes by transformation. Technical Report Research Report 2, Laboratory for Computational Stochastics, Aarhus, 1998.

[95] J.L. Jensen. Asymptotic normality of estimates in spatial point processes. *Scandinavian Journal of Statistics*, 20:97–109, 1993.

[96] J.L. Jensen and H.R. Künsch. On asymptotic normality of pseudo likelihood estimates for pairwise interaction processes. *Annals of the Institute of Statistical Mathematics*, 46:475–486, 1994.

[97] J.L. Jensen and J. Møller. Pseudolikelihood for exponential family models of spatial point processes. *The Annals of Applied Probability*, 1:445–461, 1991.

[98] O. Kallenberg. *Random measures*. Akademie Verlag, Berlin, third edition, 1983.

[99] O. Kallenberg. An informal guide to the theory of conditioning in point processes. *International Statistical Review*, 52:151–164, 1984.

[100] A.F. Karr. *Point processes and their statistical inference*. Marcel Dekker, New York, second edition, 1991.

[101] F.P. Kelly and B.D. Ripley. On Strauss's model for clustering. *Biometrika*, 63:357–360, 1976.

[102] W.S. Kendall. A spatial Markov property for nearest-neighbour Markov point processes. *Journal of Applied Probability*, 28:767–778, 1990.

[103] W.S. Kendall. Perfect simulation for the area-interaction point process. In L. Accardi and C. Heyde, editors, *Proceedings of the Symposium on Probability towards the year 2000*, Berlin, 1998. Springer-Verlag.

[104] W.S. Kendall, M.N.M. van Lieshout, and A.J. Baddeley. Quermass-interaction processes: conditions for stability. *Advances in Applied Probability (SGSA)*, 31:315–342, 1999.

[105] W.S. Kendall and J. Møller. Perfect Metropolis–Hastings simulation of locally stable spatial point processes. Technical Report 347, University of Warwick, 1999.

[106] J.F.C. Kingman. *Poisson processes*. Clarendon Press, Oxford, 1993.

[107] W. Klein. Potts-model formulation of continuum percolation.

Physical Reviews, Series B, 26:2677–2678, 1982.

[108] A.N. Kolmogorov. *Grundbegriffe der Wahrscheinlichkeitsrechnung.* Springer, Berlin, 1933.

[109] A.N. Kolmogorov. Zur Umkehrbarkeit der Statistischen Naturgesetze. *Mathematische Annalen*, 113:766–772, 1937.

[110] O.K. Kozlov. Gibbsian description of point random fields. *Theory of Probability and its Applications*, 21:339–355, 1976.

[111] K. Krickeberg. Processus ponctuels en statistique. In A. Dold and B. Eckmann, editors, *Ecole d'Eté de Probabilités de Saint-Flour X–1980*, Berlin, 1982. Springer.

[112] R.J. Kryscio and R. Saunders. On interpoint distances for planar Poisson cluster processes. *Jouranl of Applied Probability*, 20:513–528, 1983.

[113] G. Kummer and K. Matthes. Verallgemeinerung eines Satzes von Slivnyak. *Revue Roumaine de mathématiques pures et appliquées*, 15:845–870 and 1631–1642, 1970.

[114] A.B. Lawson. Discussion contribution. *Journal of the Royal Statistical Society, Series B*, 55:61–62, 1993.

[115] P.A.W. Lewis and G.S. Shedler. Simulation of non-homogeneous Poisson processes by thinning. *Naval Research Logistics Quarterly*, 26:403–413, 1979.

[116] M.N.M. van Lieshout. Stochastic annealing for nearest-neighbour point processes with application to object recognition. *Advances in Applied Probability*, 26:281–300, 1994.

[117] M.N.M. van Lieshout. *Stochastic geometry models in image analysis and spatial statistics.* CWI Tract 108, Amsterdam, 1995.

[118] M.N.M. van Lieshout. On likelihoods for Markov random sets and Boolean models. In D. Jeulin, editor, *Proceedings of the International Symposium on Advances in Theory and Applications of Random Sets*, pages 121–135, Singapore, 1997. World Scientific.

[119] M.N.M. van Lieshout. A note on the superposition of Markov point processes. Technical Report PNA-R9906, CWI, 1999. To appear in Proceedings Accuracy 2000.

[120] M.N.M. van Lieshout. Size-biased random closed sets. *Journal of Pattern Recognition*, 32:1631–1644, 1999.

[121] M.N.M. van Lieshout and A.J. Baddeley. Markov chain Monte Carlo methods for clustering of image features. In *Proceedings 5th IEE International Conference on Image Processing and its Appli-*

cations, volume 410 of *IEE Conference Publication*, pages 241–245, London, 1995. IEE Press.

[122] M.N.M. van Lieshout and A.J. Baddeley. A nonparametric measure of spatial interaction in point patterns. *Statistica Neerlandica*, 50:344–361, 1996.

[123] M.N.M. van Lieshout and A.J. Baddeley. Indices of dependence between types in multivariate point patterns. *Scandinavian Journal of Statistics*, 26:511–532, 1999.

[124] M.N.M. van Lieshout and I.S. Molchanov. Shot-noise-weighted processes: a new family of spatial point processes. *Communications in Statistics - Stochastic Models*, 14:715–734, 1998.

[125] H.W. Lotwick and B.W. Silverman. Methods for analysing spatial processes of several types of points. *Journal of the Royal Statistical Society, Series B*, 44:406–413, 1982.

[126] K.V. Mardia and G.K. Kanji, editors. *Statistics and Images*, volume 1–2, Abingdon, 1993–94. Carfax.

[127] K.V. Mardia, W. Qian, D. Shah, and K.M.A. Desouza. Deformable template recognition of multiple occluded objects. *IEEE Proceedings on Pattern Analysis and Machine Intelligence*, 19:1035–1042, 1997.

[128] E. Marinari and G. Parisi. Simulated tempering: a new Monte Carlo scheme. *Europhysics Letters*, 19:451–458, 1992.

[129] B. Matérn. Spatial variation. *Meddelanden från statens skogsforskningsinstitut*, 49:1–144, 1960.

[130] G. Matheron. *Random sets and integral geometry*. Wiley, New York, 1975.

[131] K. Matthes, J. Kerstan, and J. Mecke. *Infinitely divisible point processes*. Wiley, Chichester, 1978.

[132] K. Matthes, J. Warmuth, and J. Mecke. Bemerkungen zu einer Arbeit von X.X. Nguyen und H. Zessin. *Mathematische Nachrichten*, 88:117–127, 1979.

[133] B. McMillan. Absolutely monotone functions. *Annals of Mathematics*, 60:467–501, 1953.

[134] J. Mecke. Momentmass zufälliger Maße, 1976. Manuscript, Jena University.

[135] J. Mecke. Das n-fache Campbell–Maß, 1979. Manuscript, Jena University.

[136] K.R. Mecke. Morphological model for complex fluids. *Journal of*

Physics: Condensed Matter, 8:9663, 1996.

[137] N. Metropolis, A.W. Rosenbluth, M.N. Rosenbluth, A.H. Teller, and E. Teller. Equation of state calculations by fast computing machines. *Journal of Chemical Physics*, 21:1087–1092, 1953.

[138] S.P. Meyn and R.L. Tweedie. *Markov chains and stochastic stability*. Springer-Verlag, London, 1993.

[139] G. Mönch. Veralgemeinung eines Satzes von A. Rényi. *Studia Scientiarum Mathematicarum Hungarica*, 6:81–90, 1971.

[140] A. Mira, J. Møller, and G.O. Roberts. Perfect slice samplers. Technical Report R-99-2020, Aalborg University, 1999.

[141] J. Møller. On the rate of convergence of spatial birth-and-death processes. *Annals of the Institute of Statistical Mathematics*, 41:565–581, 1989.

[142] J. Møller. *Lectures on random Voronoi tessellations*. Springer-Verlag, Berlin, 1994.

[143] J. Møller. Markov chain Monte Carlo and spatial point processes. In O. Barndorff-Nielsen, W.S. Kendall, and M.N.M. van Lieshout, editors, *Stochastic geometry, likelihood, and computation*, London, 1999. Chapman and Hall.

[144] J. Møller and G.K. Nicholls. Perfect simulation for sample based inference. Technical Report R-99-2011, Aalborg University, 1999.

[145] I. Molchanov. *Statistics of the Boolean model for practitioners and mathematicians*. Wiley, Chichester, 1997.

[146] R. Molina and B.D. Ripley. Using spatial models as priors in astronomical image analysis. *Journal of Applied Statistics*, 16:193–206, 1989.

[147] R.A. Moyeed and Baddeley A.J. Stochastic approximation of the MLE for a spatial point pattern. *Scandinavian Journal of Statistics*, 18:39–50, 1991.

[148] D. Murdoch and P.J. Green. Exact sampling from a continous state space. *Scandinavian Journal of Statistics*, 25:438–502, 1998.

[149] D.Q. Naimann and H.P. Wynn, 1995. Personal communication.

[150] J. Neyman and E.L. Scott. Statistical approach to problems of cosmology. *Journal of the Royal Statistical Society, Series B*, 20:1–43, 1958.

[151] J. Neyman and E.L. Scott. Processes of clustering and applications. In P.A.W. Lewis, editor, *Stochastic point processes*, pages 646–681, New York, 1972. Wiley.

[152] X.X. Nguyen and H. Zessin. Integral and differential characterization of the Gibbs process. *Mathematische Nachrichten*, 88:105–115, 1979.

[153] E. Nummelin. *General irreducible Markov chains and non-negative operators*. Cambridge University Press, Cambridge, 1984.

[154] Y. Ogata and M. Tanemura. Estimation for interaction potentials of spatial point patterns through the maximum likelihood procedure. *Annals of the Institute of Statistical Mathematics*, 33:315–338, 1981.

[155] Y. Ogata and M. Tanemura. Likelihood analysis of spatial point patterns. *Journal of the Royal Statistical Society, Series B*, 46:496–518, 1984.

[156] Y. Ogata and M. Tanemura. Estimation of interaction potentials of marked spatial point patterns through the maximum likelihood method. *Biometrics*, 41:421–433, 1985.

[157] Y. Ogata and M. Tanemura. Likelihood estimation of interaction potentials and external fields of inhomogeneous spatial point patterns. In L.S. Francis, B.J.F. Manly, and F.C. Lam, editors, *Pacific Statistical Congress*, pages 150–154, Amsterdam, 1986. Elsevier.

[158] Y. Ogata and M. Tanemura. Likelihood estimation of soft-core interaction potentials for Gibbsian point patterns. *Annals of the Institute of Statistical Mathematics*, 41:583–600, 1989.

[159] S. Orey. *Limit theorems for Markov chain transition probabilities*. Van Nostrand, London, 1971.

[160] C. Palm. Intensitätsschwankungen im Fernsprechverkehr. *Ericsson Techniks*, 44:1–189, 1943.

[161] F. Papangelou. The conditional intensity of general point processes and an application to line processes. *Zeitschrift für Wahrscheinlichkeitstheorie und verwandte Gebiete*, 28:207–226, 1974.

[162] A. Penttinen. Modelling interaction in spatial point patterns: parameter estimation by the maximum likelihood method. *Jyväskylä Studies in Computer Science, Economics and Statistics*, 7:1–105, 1984.

[163] P. Peskun. Optimum Monte Carlo sampling using Markov chains. *Biometrika*, 60:607–612, 1973.

[164] D.B. Philips and A.F.M. Smith. Dynamic image analysis using

Bayesian shape and texture models. In K.V. Mardia and G.K. Kanji, editors, *Statistics and Images Volume 1, Advances in Applied Statistics, a supplement to Journal of Applied Statistics Volume 20*, pages 299–322, Abingdon, 1993. Carfax.

[165] C.J. Preston. *Random fields*. Springer, Berlin, 1976.

[166] C.J. Preston. Spatial birth-and-death processes. *Bulletin of the International Statistical Institute*, 46:371–391, 1977.

[167] J.G. Propp and D.B. Wilson. Exact sampling with coupled markov chains and applications to statistical mechanics. *Random Structures and Algorithms*, 9:223–252, 1996.

[168] R.-D. Reiss. *A course on point processes*. Springer, New York, 1993.

[169] A. Rényi. A characterisation of Poisson processes (English translation of original in Hungarian). In P. Turan, editor, *Selected papers of Alfred Rényi*, volume 1, pages 622–628, 1956/1976.

[170] B.D. Ripley. The foundations of stochastic geometry. *The Annals of Probability*, 4:995–998, 1976.

[171] B.D. Ripley. Locally finite random sets: foundations for point process theory. *The Annals of Probability*, 4:983–994, 1976.

[172] B.D. Ripley. On stationarity and superposition of point processes. *The Annals of Probability*, 4:999–1005, 1976.

[173] B.D. Ripley. The second-order analysis of stationary point processes. *Journal of Applied Probability*, 13:255–266, 1976.

[174] B.D. Ripley. Modelling spatial patterns. *Journal of the Royal Statistical Society, Series B*, 39:172–192, 1977.

[175] B.D. Ripley. Tests of 'randomness' for spatial point patterns. *Journal of the Royal Statistical Society, Series B*, 41:368–374, 1979.

[176] B.D. Ripley. *Spatial statistics*. Wiley, Chichester, 1981.

[177] B.D. Ripley. Statistics, images and pattern recognition. *Canadian Journal of Statistics*, 14:83–111, 1986.

[178] B.D. Ripley. *Stochastic simulation*. Wiley, Chichester, 1987.

[179] B.D. Ripley. *Statistical inference for spatial processes*. Cambridge University Press, Cambridge, 1988.

[180] B.D. Ripley. Gibbsian interaction models. In D.A. Griffiths, editor, *Spatial statistics: past, present and future*, pages 1–19, New York, 1989. Image.

[181] B.D. Ripley and F.P. Kelly. Markov point processes. *Journal of the London Mathematical Society*, 15:188–192, 1977.

[182] B.D. Ripley and A.I. Sutherland. Finding spiral structures in images of galaxies. *Philosophical Transactions of the Royal Society of London, Series A*, 332:477–485, 1990.

[183] B.D. Ripley and C.C. Taylor. Pattern recognition. *Science in Progress*, 71:413–428, 1987.

[184] H.E. Robbins. On the measure of a random set. *Annals of Mathematical Statistics*, 15:70–74, 1944.

[185] H.E. Robbins. On the measure of a random set II. *Annals of Mathematical Statistics*, 16:342–347, 1945.

[186] J.S. Rowlinson. Penetrable sphere models of liquid-vapor oquilibrium. *Advances in Chemical Physics*, 41:1–57, 1980.

[187] J.S. Rowlinson. Probability densities for some one-dimensional problems in statistical mechanics. In G.R. Grimmett and D.J.A Welsh, editors, *Disorder in physical systems*, pages 261–276, Oxford, 1990. Clarendon Press.

[188] H. Rue. New loss functions in Bayesian imaging. *Journal of the American Statistical Association*, 90:900–908, 1995.

[189] H. Rue and M.A. Hurn. Bayesian object identification. *Biometrika*, To appear.

[190] H. Rue and A.R. Syversveen. Bayesian object recognition with Baddeley's Delta loss. *Advances in Applied Probability (SGSA)*, 30:64–84, 1998.

[191] D. Ruelle. *Statistical mechanics*. Benjamin, New York, 1969.

[192] L.A. Santalo. On the first two moments of the measure of the random set. *Annals of Mathematical Statistics*, 18:37–49, 1947.

[193] R. Saunders, R.J. Kryscio, and G.M. Funk. Poisson limits for a hard-core clustering model. *Stochastic Processes and Applications*, 12:97–106, 1982.

[194] K. Schladitz and A.J. Baddeley. A third order point process characteristic. Technical Report 1997/20, University of Western Australia, 1997.

[195] J. Serra. *Image analysis and mathematical morphology*. Academic Press, London, 1982.

[196] A. Särkkä. *Pseudo–likelihood approach for pair potential estimation of Gibbs processes*. PhD thesis, University of Jyväskylä, 1993.

[197] A. Särkkä and H. Högmander. Multitype spatial point patterns with hierarchical interactions. Technical Report 1998:1, Gothen-

burg University, 1998.

[198] Y. Shreider, editor. *Method of statistical testing, Monte Carlo method*, Amsterdam, 1964. Elsevier.

[199] B.W. Silverman. *Density estimation*. Chapman and Hall, London, 1986.

[200] A. Simó. *Modelización y cuantificación de dependicias en procesos puntuales y conjuntos aleatorios bivariantes*. PhD thesis, University of Valencia, 1995.

[201] K. Sivakumar. *Morphological analysis of random fields: theory and applications*. PhD thesis, Johns Hopkins University, 1997.

[202] K. Sivakumar and J. Goutsias. Morphologically constrained discrete random sets. In D. Jeulin, editor, *Proceedings of the International Symposium on Advances in Theory and Applications of Random Sets*, pages 121–135, Singapore, 1997. World Scientific.

[203] I.M. Sobol. *Die Monte Carlo Methode*. Deutscher Verlag der Wissenschaften, Berlin, 1971.

[204] A. Sokal. Monte Carlo methods in statistical mechanics: foundations and new algorithms, 1989. Course de troisième cycle de la physique en Suisse romande.

[205] H. Solomon. Distribution of the measure of a random two-dimensional set. *Annals of Mathematical Statistics*, 24:650–656, 1953.

[206] A. Stein, P. Droogers, and H. Booltink. Point processes and random sets for analyzing patterns of methylene blue coloured soil. *Soil and Tillage Research*, 46:273–288, 1998.

[207] A. Stein, M.N.M. van Lieshout, and H.W.G Booltink. Spatial interaction of blue-coloured soil stains. Technical report, Wageningen Agricultural University, 1999.

[208] D. Stoyan, W.S. Kendall, and J. Mecke. *Stochastic geometry and its applications*. Wiley, Chichester, second edition, 1995.

[209] D. Stoyan and H. Stoyan. *Fractals, random shapes and point fields*. Wiley, Chichester, 1994.

[210] D.J. Strauss. A model for clustering. *Biometrika*, 62:467–475, 1975.

[211] R.H. Swendsen and J.-S. Wang. Nonuniversal critical dynamics in Monte Carlo simulations. *Physical Review Letters*, 58:86–88, 1987.

[212] R. Takacs. Estimator for the pair-potential of a Gibbsian point process. Technical Report 238, Institut für Mathematik, Johannes

Kepler Universität Linz, 1983.

[213] R. Takacs. Estimator for the pair potential of a Gibbsian point process. *Statistics*, 17:429–433, 1986.

[214] R. Takacs and T. Fiksel. Interaction pair-potentials for a system of ants' nests. *Biometrical Journal*, 28:1007–1013, 1986.

[215] M. Talbort, editor. *Quantitative methods for sustainable agriculture*, Edinburgh, 1994. Scottish Agricultural Statistics Service.

[216] E. Thőnnes. Perfect simulation of some point processes for the impatient user. *Advances in Applied Probability (SGSA)*, To appear.

[217] C.J. Thompson. *Mathematical statistical mechanics*. Macmillan, New York, 1976.

[218] L. Tierney. Markov chains for exploring posterior distributions. *Annals of statistics*, 22:1701–1728, 1994.

[219] G. Upton and B. Fingleton. *Spatial data analysis by example*. Wiley, Chichester, 1985.

[220] O. Wälder and D. Stoyan. Models of markings and thinnings of Poisson processes. *Statistics*, 29:179–202, 1997.

[221] B. Widom and J.S. Rowlinson. A new model for the study of liquid-vapor phase transitions. *Journal of Chemical Physics*, pages 1670–1684, 1970.

Index